Singapore's Permanent Territorial Revolution

Singapore's Permanent Territorial Revolution

Fifty Years in Fifty Maps

Rodolphe De Koninck

Cartography by
Pham Thanh Hai & Marc Girard

NUS PRESS
SINGAPORE

© 2017 Rodolphe De Koninck

Published by:
NUS Press
National University of Singapore
AS3-01-02, 3 Arts Link
Singapore 117569

Fax: (65) 6774-0652
E-mail: nusbooks@nus.edu.sg
Website: http://nuspress.nus.edu.sg

ISBN: 978-981-4722-35-3 (casebound)

National Library Board, Singapore Cataloguing in Publication Data

Names: Koninck, Rodolphe De. | Pham Thanh Hai, 1969- cartographer. | Girard, Marc, 1964- cartographer.
Title: Singapore's permanent territorial revolution: fifty years in fifty maps / Rodolphe De Koninck,
        Pham Thanh Hai and Marc Girard.
Description: Singapore: NUS Press, [2017] | Includes bibliographic references and index.
Identifiers: OCN 962466310 | 978-981-47-2235-3 (casebound)
Subjects: LCSH: Land use--Singapore--Planning--Maps. | Singapore--Social conditions--Maps. | Singapore--
        Economic conditions--Maps. | Singapore--History--Maps.
Classification: DDC 912.5957--dc23

Designed by: Nelani Jinadasa
Printed by: Markono Print Media Pte Ltd

Contents

Preface

The preparation of this essay, which takes the form of an atlas, is the result of a collaborative endeavour. It also has a rather long history, having gone through three distinct phases over the last quarter century. In 1990–91, with the help of several persons and in particular Guy Dorval, Louise Marcotte and Andrée Lavoie at Laval University, I prepared a bilingual atlas that appeared in late 1992 in France, under the title *Singapour. Un atlas de la révolution du territoire / Singapore. An Atlas of the Revolution of Territory.* Nearly 15 years later, Singaporean friends convinced me to update and augment the atlas and submit it to NUS Press. With the help of Julie Drolet and Marc Girard, who both handled the cartography, the work was done in 2005 and 2006. The new manuscript was accepted by NUS Press which published it in 2008 under the title *Singapore: An Atlas of Perpetual Territorial Transformation.*

It then looked as if this was the end of the Singapore Atlas adventure! But because of my unwavering fascination with Singapore and, once again following the prompting of Singaporean friends, including newly made ones, the decision was taken in August 2014 to prepare the present third atlas. This time the maps were to be done by Marc Girard, who was soon joined by Pham Thanh Hai. Both had to work very hard, relentlessly, as this new version was to contain 50 plates – instead of the 34 that made up the 2008 atlas – while nearly all would have to be more elaborately commented, a task which was my responsibility.

Over the course of nearly two years, as each of us was involved in several other ventures, we still managed to get together frequently, in person or online, in order to work on the atlas, to debate its contents, the validity of the sources available, as well as the intricacies of an analysis relying on time series representations. These were the same problems that had to be dealt during the preparation of the two previous atlases and particularly the 2008 one. Singapore changes so rapidly, yet in such a generally well-monitored fashion, that we frequently had doubts about the need for, as well as the validity of, an analytical atlas. Nevertheless, thanks to encouragement from friends and colleagues, particularly in

Singapore and in Montreal, two thriving yet vastly different cities half way around the world from one other, we completed the task. For this we are grateful to many, particularly in those two places, more specifically at the Université de Montréal and at the National University of Singapore (NUS). At NUS Press itself we were fortunate to be able to count on the help and critical comments and at times direct input from several persons, in particular Christine Chong, Peter Schoppert and Nelani D. D. Jinadasa.

Also in Singapore, several other persons generously provided exceptional help and advice. I wish to thank three in particular. The first, Patrick Low, is a friend from the days when I was undergoing my PhD at the University of Singapore (1967–70). After we had lost touch with one another for several decades, we reconnected a few years ago, at a time when the 2008 atlas had already been published. Since then, Patrick has become once again a keen interlocutor, in his capacity as a concerned Singaporean, very knowledgeable about his Island Republic, which he has helped me tour all over once again, except this time we came across hardly any dirt roads!

The second exceptional interlocutor, and those who know him will not be surprised to read his name here, is Tay Kheng Soon. Also a friend from the late 1960s, Kheng Soon is an architect, a very creative one, an intellectual always full of ideas, and a formidable debater when it comes to Singaporean architecture, geography and urban planning. I have managed to meet with him, if at times only too briefly, on almost every one of my near yearly visits to Singapore over the last half century! And each time, our discussions and his unique combination of intellectual curiosity, open-mindedness and creativity have been an inspiration to me.

The third outstanding interlocutor is Mok Ly Yng, Mok to friends, whom I only got to know in August 2015. Since then, his help has become invaluable. Mok turned out to be not only a very well-informed and concerned Singaporean, but also one with a vast knowledge of anything and everything having to do with maps, especially those which concern Singapore, past and present! Over the last few months, he provided us with plenty of data and advice. Without his input, this atlas would be much poorer.

As implied above, beyond these three very helpful friends are many others, some of which have equally contributed directly or indirectly to the production of this work. Most belong to the category of old friends. These include Willy Lim and Lena Lim, Chan Heng Chee, Chua Beng Huat, Wong Poh Poh, Julie Yeo, Irene Chee, Lim Kim Leng, Paul Kratoska, Victor Savage and John Miksic. I would also like to mention James Sidaway for pointing out some little known sources, Kim Ick-Hoi (Rick) for his useful advice on map sources, Ong Ai Bin for data on the Association of Banks in Singapore and Theresa Wong who, during my July–August 2015 sojourn in Singapore, insisted I should meet with Mok Ly Yng. How right she was!

During that two-month spell in Singapore, I was hosted by the Asia Research Institute at NUS as a Senior Visiting Research Fellow. That opportunity allowed me to collect much information and many documents and, just as important, to relive in Singapore, enjoy the company of Singaporean friends cum counsellors and, may I add,

share their passion for gastronomy! As a person of French culture, I realise how important that is, including for intellectual exchange!

I also wish to thank the Groupement d'Intérêt Public (GIP) Reclus, in Montpellier, France, which published the 1992 atlas. The GIP gracefully granted us permission to make use of the original map files, as did the Éditions Belin in Paris for the map files from my 2006 monograph on Singapore. Several of these were adapted to produce plates that appeared in the 2008 atlas published by NUS and, of course, readapted for use in the present one. I equally express my gratitude to the Maison Suger in Paris, where I have resided several times over recent years. During some of these sojourns, I was able to work on my Singaporean material and to benefit from discussions with French colleagues, in particular Charles Goldblum. Back in Montreal, Bernard Bernier and Bruno Thibert were also very helpful.

Finally, Hai and Marc join me in expressing our gratitude to the Social Sciences and Humanities Research Council of Canada for its financial support to the Canada Chair of Asian Research, of which I am the holder.

Rodolphe De Koninck
Paris, May 2016

Introduction: The Territorial Hypothesis

"At times the contemplation of the attitudes of the people of Singapore drives one to tears, more often it leaves one in a state of stunned and slightly resentful admiration."

– D. J. Enright, 1969, p. 181

An observer of the Singaporean scene cannot remain indifferent: he is either fascinated or shocked, never unconcerned. Weary or still curious Asia scholars, disillusioned or militant development specialists, ignorant or well-informed tourists, all are impressed by the dynamism of the small island republic, by what is done away with as well as by what is achieved in the midst of what appears to be a perpetual transformation. Indeed, for well over 50 years now – and particularly since 1965, when it became a fully independent city-state – Singapore has been an effervescent laboratory of economic, social and environmental transformation and innovation, with its political rulers appearing remarkably stable, even conservative, and its inhabitants apparently acquiescent.

Studies dealing with these changes are quite numerous. Occasionally carried out by foreigners, they are more frequently authored by local citizens, particularly civil servants, including academics. They are generally published in the form of readers, and, while mentioning the magnitude of the environmental transformations, seem to consider the changes an inevitable consequence of the Singaporean nation-building process. The reasoning goes somewhat as follows: the island of Singapore being small (currently about 720 square km), the needs and aspirations of its citizens being great, the former must be thoroughly transformed and extended.

Legitimate in itself, this interpretation – which Singapore authorities endorse– leaves no room for the following hypothesis: the remarkable and much-vaunted efficiency in the implementation of socio-economic transformations, even more so the resignation with which the Singaporean population seems to live through them, are at least partly attributable to the permanent transformation of its living space. According to this hypothesis (formulated more fully in De Koninck, 1990 and 1992), the manipulation of the environment and the repeated erosion or ephemeral character of all spatial bearings at the local level allow for only one level of territorial allegiance: that of the Republic of Singapore. The constant redefinition of these spatial and environmental bearings, while associated with other forms of monitoring, and not necessarily the result of a concerted decision, is not a mere consequence of changes accomplished in the political, economic and social spheres, but a tool. Spatial instability and territorial alienation at the local level foster social docility or at least assent. Short of relying on a massive survey of Singaporeans' topophilia – from the Greek words topos for place and philia for love, hence literally meaning love of place or sense of place – such a hypothesis would be difficult to verify. But its validity can at least be emphasized by an illustration of the nature and magnitude of the permanent transformations that seem to have become an inevitable feature of the Singaporean landscape since the late 1950s and even more in the mid-1960s.

I may add that the territorial hypothesis does not result from a fortuitous hunch, but rather from observations made nearly 50 years ago when I was pursuing my PhD at the University of Singapore. For this thesis research (1967–70), which dealt with the Chinese market gardeners then still active in the Singaporean countryside, I roamed extensively across the entire island and became somewhat aware of the transformation that was in the making (De Koninck, 1970, 1975).

But it was only after several additional and mostly brief sojourns, performed between 1972 and 1989, when Singapore no longer represented for me a privileged object of study, that the magnitude of the changes really began to bewilder me. Losing or having difficulty finding my way in an environment that had once been so familiar brought the following questions to my mind. How can such a pace be maintained? How do the citizens of this country, who possess such rich cultural backgrounds, accept the fact that the very ground on which they live is perpetually changing, that the rug is constantly pulled from under their feet? How – notwithstanding the obvious improvements in their standards of living – do they cope with the near-constant removal or transformation of their landmarks? All the answers to my questions, all the studies consulted remained unsatisfactory. Then a new interrogation took form: somewhat like the case of the earth's revolutions around its axis, rapid and definitive to the point where most people do not attempt to understand its movement, aren't average Singaporeans so caught up in such a whirlwind that they have no choice but to hang on ... or drop off?

To further substantiate this interrogation, documentary as well as field research was carried out in Singapore in 1990. As mentioned in the preface, it led to the production of a first edition of this atlas, essentially dedicated to the illustration of the whirling territorial redistributions. Published in France in 1992 in a bilingual version (French and English), that initial book is now out of print. But Singapore's way of doing things has not really changed. On the contrary, the perpetual transformation of the environment, in any form – physical, urban, rural, residential, infrastructural, cultural, etc. – is still proceeding apace. In other words, the territorial hypothesis still seems applicable. The permanent overhaul of the Singaporean environment – whether one calls it development, improvement or upgrading – still seems to be a way of life, or rather a way of managing a country, which

some may call governance. But now the country is definitely prosperous and therefore one where constant and rapid territorial transformation does not appear so indispensable, so obviously urgent.

Hence this thoroughly revised and updated edition of the 1992 atlas and of the much more elaborate version published by NUS Press in 2008. I still hypothesize that the systematic overhaul of the Singaporean environment represents a deliberate and politically motivated form of social transformation and social management, a transformation monitored from above. I however do not claim that all the results are wholly intentional, that by practising a policy of tabula rasa (Koolhas, 1995), or table rase as I wrote in 1990, Singaporean authorities intend to destabilize Singaporean residents. That is not necessarily true. But the result remains, or so I hypothesize: citizen topophilia is largely done away with or at least constantly redefined, thus ensuring that the state continues to exercise total control.

In 2008 with the help of Julie Drolet and Marc Girard, and presently with Pham Thanh Hai and Marc Girard once again, we were and are able to further document and illustrate the validity of this territorial hypothesis. Following the examination of numerous forms of apparently systematic and permanent territorial upheavals, I conclude with a spatial model "par excellence": that of a State in perpetually planned and replanned transformation, constantly having to adapt itself to models of its own making.

Four additional basic comments are necessary here. First, this atlas is neither meant to be comprehensive nor to act as a reference work. Rather, it constitutes an essay, an attempt to consolidate a hypothesis relying on an original approach and making use of relatively broad and primarily cartographic source material. Second, just like the atlas itself, the individual maps along with their comments are not meant to be complete, whether about industry, housing or tourism or overall globalisation processes, all topics on which statistics quickly become outdated in Singapore; nor are they meant to cover the entire Singaporean "project" and its underlying social and political philosophy, including that of "Singapore Inc.". The choice of topics and manner of their handling were largely determined by their potential contribution to the illustration of the territorial hypothesis. Third, several potentially interesting themes could not be addressed. In some cases, such as foreign workers to which more than one plate should have been dedicated, this was because of the political sensitivity of the issue and paucity of data available. In several other cases, we were also faced with a lack of data, particularly comparable data allowing for the representation of time series. Fourth, for the themes retained, available data did not always concern identical years, although, for most of them we did try and favour a diachronic and even more a longitudinal representation, covering three or even four dates.

It remains quite evident that an attempt to comprehend the permanently animated Singaporean scene could rely on much more elaborate illustrations, at a variety of scales, including that of neighbourhoods and even streets, at least those that can still be retraced. Nevertheless, the illustrations that follow do suggest some keys.

This brings me to the question of the title of the book. As in the initial 1992 atlas, I have chosen to return to the word "revolution" which, to most people, would seem far removed from Singapore. Of course, Singapore is not revolutionary in the classical social or political sense. No open physical violence is reverted to and, of course, neither social class nor any individual is being overthrown here. Yet, as exemplified by the constant transformation of the physical environment, and given the consequences it has on the citizenry, I still consider the concept of revolution, of permanent revolution, to be applicable. The suggestion – present throughout the atlas – is that in Singapore, this permanent revolution has become a form of governance, allowing for better and permanent control.

Finally, since the title says 50 years in 50 maps, why is it, some may ask, that so many of these maps do not cover exactly the period 1965 to 2015? Indeed, many of the sequences refer to a longer period, for example 1957 to 2015 or 1958 or 1959 or 1960 to 2015. Here again this is largely attributable to questions of data availability and comparability. The first census after Singapore's 1965 independence was carried out in 1970 and the last one prior to it in 1957. Also, in 1958, topographical maps of the island were produced at 1:25,000 (four sheets) and 1:63,000 (one sheet). Along with the 1957 census these represented the most useful sources for our mapping purposes and they were used for the initial 1992 edition of the atlas. Attempting to cover exactly the 50-year period 1965–2015 would have meant starting from scratch and relying on sources much less reliable than a census and topographical maps.

NOTE: The international boundaries in this atlas are not authoritative. They are depicted for illustrative purposes only.

Chapter 1
Setting the Stage

"This is by far the most important station in the East; and, as far as naval superiority and commercial interests are concerned, of much higher value than whole continents of territory."

– Stamford Raffles, 1819 (in Turnbull, 1989, p. 12)

"From age to age, a port had arisen in this part of the Malay world. (…) In the past, such cities had been dominated by Malay rulers and the maritime peoples of the region. In the nineteenth century, even though it was under British rule, Singapore shared fully in the trans-Asian, maritime trading culture that had a heritage of over fifteen centuries."

– Carl A. Trocki, 2006, p. 9

"Early modern cartographers understood the island of Singapore not so much as a convenient spot in the maritime artery connecting long-distance trade between east and west, but rather as a critical location at a notional divide between the greater Bay of Bengal and the South China Sea, with the straits acting as a "gateway" leading from one maritime sphere to another."

– Borschberg, 2015, p. 28

Well before Western navigators and empire builders came to Southeast Asia, Singapore had been playing a key role in trading networks within its immediate surroundings as well as beyond, all the way to India on one side and China on the other.

So, when the British asserted their control over Singapore in 1819 – thanks to Stamford Raffles, who, in the name of the East India Company, established a soon thriving trading post halfway between Calcutta, the capital of the British Indian Empire, and Canton, the main gate to China – they were doing so in a well-known location. Except that, given their colonial ambitions, as early as 1823, even before the conclusion of the 1824 Treaty of London granting Melaka to the British, the island's population had grown from a mere few hundreds to more than 10,000. The opening of the Suez Canal in 1869 consolidated Singapore's trading function in the service of British colonial interests. A free port and emporium surrounded by resource-rich territories, Singapore was both a haven and a transit point for immigrants. Its destiny was associated with that of the Malay Peninsula, where tin and eventually rubber were produced in vast quantities. The opening of a large naval base in 1938 was followed shortly by the establishment of two Royal Air Force bases.

That did not prevent "Fortress Singapore" from collapsing under the Japanese attack of February 1942. In 1945, after more than three years of occupation, demographic growth resumed. Overcrowding of the urban core and major suburbs became acute, with one third of the population living in slums. The situation had become explosive by the end of the 1950s, as poverty, crime, social conflicts and political unrest increased. Solutions had to be found. Among these, the territorial revolution soon became an essential one.

An aerial view of the Singapore River and the surrounding Central Business District (CBD) in 1971. Bumboats are moored along the sides of the Singapore River. From the Kouo Shang-Wei Collection 郭尚慰收集. All rights reserved, Family of Kouo Shang-Wei and National Library Board Singapore 2016.

1 Singapore: A Global Strategic Location

"The Gibraltar of the East... the gateway to the Orient... the bastion of British might."

– Sydney Morning Herald, 14 February 1938[1]

It has almost become a truism to refer to Singapore as being "strategically located". Yet, truism or not, among the major cities on the planet, Singapore is indeed one of the most strategically located. And this applies at several scales. Firstly, as it nearly straddles the Equator and is located at the junction of the Pacific and Indian Oceans, it stands in a nodal position on the sea routes linking Europe with the Far East as well as South Asia to East Asia, via the so-called China Road. Such a location became particularly strategic after the opening of the Suez Canal in 1869, at a time when British imperial power in the Far East was rising, which led to an even higher level impact. Secondly, Singapore's advantageous position has been gradually confirmed at the scale of the world, itself involved in an endless process of trade globalisation. The contemporary web of Asian shipping routes clearly shows how Singapore is centrally located within that world, and particularly within Eurasia. The port of Singapore stands at the core of a network of some 200 shipping lines, linking over 600 ports located in more than 120 countries.

Obviously, Singapore has evolved way beyond its status as a military fortress within a flailing British empire. To its function as a commanding stopover on the world's increasingly busiest sea route – and also as its busiest port – has recently been added its role as an airline hub between

Global Shipping Routes

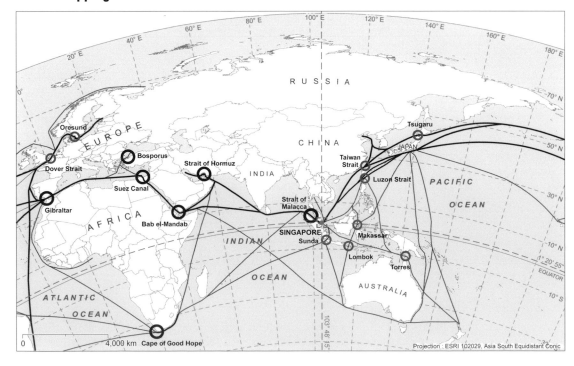

Main Maritime Shipping Routes
— Core Route
— Secondary Route

Main Strategic Maritime Passages
⊙ Primary Chokepoint
○ Secondary Chokepoint

Eurasia and Australasia (p. 112), while its central position on the maritime routes linking Australasia as well as Southeast and East Asia with Southern Africa has also been enhanced (p. 8). Not only has the Island Republic enhanced its status as a trading hub, it has also become a centre of innovation in several fields, from housing and urban planning to, among other things, petrochemical industry, biotechnology and medical sciences, information technology, finances and higher learning.

---

1   Turnbull, 2009, p. 171.

Southeast Asian Shipping Routes

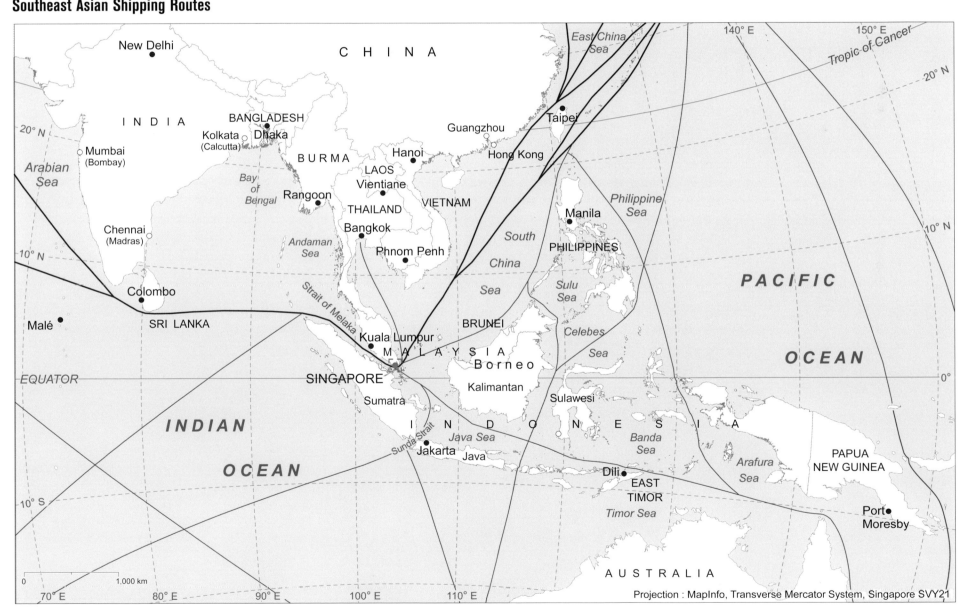

2 Singapore: A Regional Strategic Location

"Singapore's location at the south end of the Straits attracted attention from the two main empires of the fourteenth century, one based on the mainland, the other in the islands. This position underscores both Singapore's historical potential and its vulnerability, which are two sides of the same coin."

– John Miksic, 2013, p. 185

This statement by a leading historian of premodern Singapore, concerning its historical potential and its vulnerability, rejoins ex-Prime Minister Lee Kwan Yew's often-repeated opinion that Singapore had an immense potential but was also vulnerable, the two sides of the same coin being largely related to its strategic position.

For not only is Singapore located strategically on the world map, it is also within Asia and, even more narrowly, Southeast Asia. To begin with, as it stands at the very tip of the Malay Peninsula, it constitutes the ultimate southern projection of the Asian continent. More fundamentally, it occupies a commanding position at the southern extremity of the Strait of Malacca. And its micro location happens to be in calm equatorial and well-sheltered waters, a fundamental asset for what has become the world's busiest sea port and number one bunkering harbour. And, as already mentioned, this position is on the key passage between the Indian and Pacific oceans, halfway between India and China, at the heart of what is still the China-Road and, increasingly significant, the world's busiest sea passage.

While Singapore has for centuries benefitted from this strategic situation, its own initiatives – or rather those of its inhabitants – have consolidated it, particularly over the last half century. The choice to evolve from depending nearly exclusively on the status of trade emporium to diversifying into that of an innovation hub has been enhanced in several fields of endeavour, such as, of course, port and airport operation, urban planning and management, housing, education, industrial development and financial services, among others. As a consequence, Singapore's position as a global hub has been more than consolidated. It has made itself increasingly indispensable to the world.

On 23 May 2008, the International Court of Justice awarded sovereignty over the island of Pedra Branca to Singapore, and a nearby feature, the Middle Rocks, to Malaysia: the Horsburgh Lighthouse was built on Pedra Branca in 1851 to aid ships entering the Straits of Singapore from the South China Sea.
Map courtesy of Quiex, on the Wikimedia Commons.

TOP: Singapore in Asia, between the Indian and Pacific Oceans.

BOTTOM: Singapore in Southeast Asia, in the middle of the South China Sea and the Java Sea.

UTM projection (WGS84) 48 N area

● Important city   ● Capital   ▨ Zone where the scale is application

3 Singapore in the Midst of Historical Trade Centres

"[Raffles'] conviction that Singapore had experienced a period of prominence was reinforced soon after the British took possession of the island."

– John Miksic, 2004, p. 45

Situated some 140 km north of the Equator, the island of Singapore occupies a sheltered position at the southern end of the Malay Peninsula and of the Strait of Malacca. Throughout history, for reasons related as much to navigation and settlement as to the policies of empires, this passage became the most important link between the Indian and Pacific Oceans and a key one in the world. Nevertheless, even if ships had long sailed through the Straits, Singapore began to play a more eminent role only after the founding of the British trading post in 1819 by Stamford Raffles. It then quickly became an essential component of the British Empire and its trading network with the Far East.

But Singapore's earlier history, the richness of which has been increasingly recognised over the last two decades, had been overshadowed by that of various states and kingdoms having exercised some form of hegemony in the region: in particular Srivijaya, Siam and the sultanates of Aceh, Melaka and Johor. Thanks to nearly three decades of excavations (1984–2012) and further scrutiny of written documents, archaeological and historical findings confirm that the island had probably been the seat of a Malay kingdom from the end of the thirteenth century until the early fifteenth. It had, in 1320, hosted an envoy from the Yuan Emperor. Under the names of "Temasek" and "Singapura" (Lion City), in use from the end of the 14th century, it was subsequently known as a refuge for pirates, like many surrounding archipelagos, islands and estuaries. But much more than that, it should be remembered as one of the early urban centres in Southeast Asia, a port that had, particularly during the fourteenth century, hosted traders from other parts of Asia, "creating a multiethnic population similar to that which developed after Raffles revived the ancient port" (Miksic, 2013, p. 20).

A closer look at the maze of islands between Sumatra and the Malay Peninsula reveals the quality of the island of Singapore's sheltered location, arguably superior to those of other historical trade centres found in the region, whether in the nearby Riau Archipelago or on the coast and estuaries of the Malay Peninsula or of Sumatra Island. The virtues of that location had obviously been recognised by local populations centuries before the arrival of the Europeans in the region.

In front of the old Supreme Court, backhoe operations confirmed the presence of fourteenth-century artifacts at the Colombo Court archaeological dig. Reprinted with permission from *Singapore and the Silk Road of the Sea*, NUS Press, 2013.

Singapore in the midst of its historical trade centres. The historical names or spellings are in parentheses below the contemporary names.

4 Singapore as a Migrants' Haven

"The history and fortunes of Singapore […] have always been closely intertwined with migration."
– Yeoh and Lin, 2012

The demographic composition of Singapore's entire population – including citizens, permanent residents and foreigners on a temporary work permit – testifies to its long-standing history as a migrant's haven. Of course, this notion of a "haven" must be qualified. To be fair, even now not all immigrants have an easy adaptation. But no one will deny that the island's history, that of its premodern as well as modern settlement, of its development as a world-class hub and, finally, of its current economic growth and predicaments are all closely linked to immigration: Singapore has long been a privileged destination for people looking for employment. While such a statement may not be easy to substantiate for Singapore's pre-nineteenth century history, it is clearly the case for its development under British colonial rule and perhaps even more since its independence in 1965.

When the British took possession of Singapore in 1819, it was inhabited by fewer than a thousand people, mostly Malays, a few tens of Chinese as well as an undetermined number of Orang Laut. The latter were sea people who preferred to live on their boats, stationed in one or the other of the many of the island's river estuaries, particularly those draining into the narrow Strait of Johor. Under the leadership of Stamford Raffles, the British were intent on developing rapidly their new settlement and to make it a trading emporium. On both sides of the Singapore River estuary, town plans were rapidly implemented to host new migrants in clearly laid out ethnic neighbourhoods.

Malay and Chinese migrants were the first to arrive in large numbers, mostly from the Malay Peninsula, the Riau islands and Sumatra, but also, particularly in the case of the Chinese, from places as far as Java or Bangkok. By 1823, the population had reached nearly 11,000, the Malays representing the principal group, ahead of the Chinese. Ten years later, when the population had nearly doubled, the latter had taken the lead. They have never lost it since and have in fact increased it substantially. By 1833, the number of Indian migrants had also risen to a point where they accounted for 11 per cent of the total population.

Immigration has continued more or less unabated, except during the Japanese occupation years (1942–45). It also slowed down considerably during the 1950s and 1960s when Singapore's political status was gradually transformed.

By the 1980s, the island Republic's economy had reached such a rate of growth that demand for foreign labour began to rapidly rise once again. But the nature of the increase was somewhat different: what the city-state needed was not so much more population as a larger workforce. The result has been an influx of foreign workers, the vast majority being allowed to reside in Singapore only for the duration of their

By the 1980s the nature of the increase was somewhat different: what the city-state needed was not so much more population as a larger workforce.

Share of Foreign Workers in Labour Force, 1970–2014

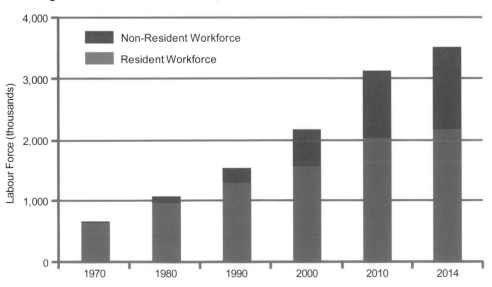

Main Migration Flows to Singapore since 1819

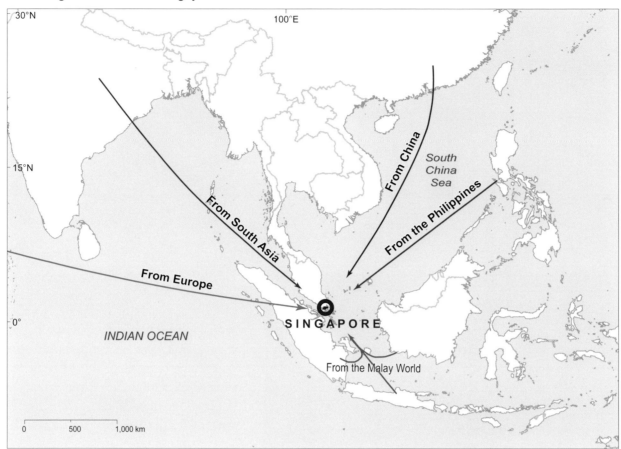

TOP: From the colonial period to today, Singapore's migrant population has originated primarily from China, the Malay World, the Philippines, South Asia and Europe.

BOTTOM: Singapore's population growth from 1819 to 1833.

Population Growth between 1819 and 1833

	1819	1823		1833	
	Inhab.	Inhab.	%	Inhab	%
Europeans		74	1	119	1
Indians		756	7	2,315	11
Chinese	Mostly Malays, Orang Laut, a few Chinese	3,317	31	8,517	41
Malays		4,580	43	7,131	34
Others		1,956	18	2,798	13
Total	Less than 1,000	10,683	100	20,880	100

A Changing Population Structure, 1970–2014

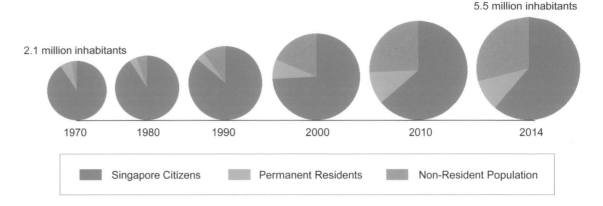

5.5 million inhabitants

2.1 million inhabitants

1970   1980   1990   2000   2010   2014

■ Singapore Citizens   ■ Permanent Residents   ■ Non-Resident Population

TOP: Singapore's ever growing population, and its relative proportion of citizens, permanent residents, and non-residents.

BELOW: Raffles Place, which used to be Singapore's "commercial square" before it became the heart of the financial district.

work contract, generally no longer than two or three years. As a consequence, non-residents (i.e., those who are neither citizens nor permanent residents) now account for more than 35 per cent of Singapore's workforce, while the proportion of its overall population represented by non-residents has risen from some three per cent in 1970 to ten per cent in 1990, 26 per cent in 2010 and 29 per cent in 2014! (p. 44).

Over the years, the Singapore government has frequently adjusted its policy regarding work permits granted to foreigners, whether low-skilled workers or high-skilled employment pass holders, the latter often eligible to eventually apply for permanent residence. At times, particularly when protests from residents have become more vocal over a number of issues regarding "foreigners", some rules have been tightened. But Singapore's immigration policies have remained those of a country that needs to compensate for an aging resident population, whose members are generally unwilling to work as manual labourers. Furthermore, it is a country also intent on expanding its economy and its territory! One consequence is that, with an in migration rate standing at 14.55 in 2014, Singapore occupies the eighth rank in the world, way ahead of all other highly developed countries.

5 Contemporary Singapore

While Singapore was never really mountainous, it was and still remains far from flat, with several ranges of hills structuring the territory, particularly at its geographical core. This renders even more astonishing the extensive transformations administered, so to speak, to the island over the last five decades.

Topography

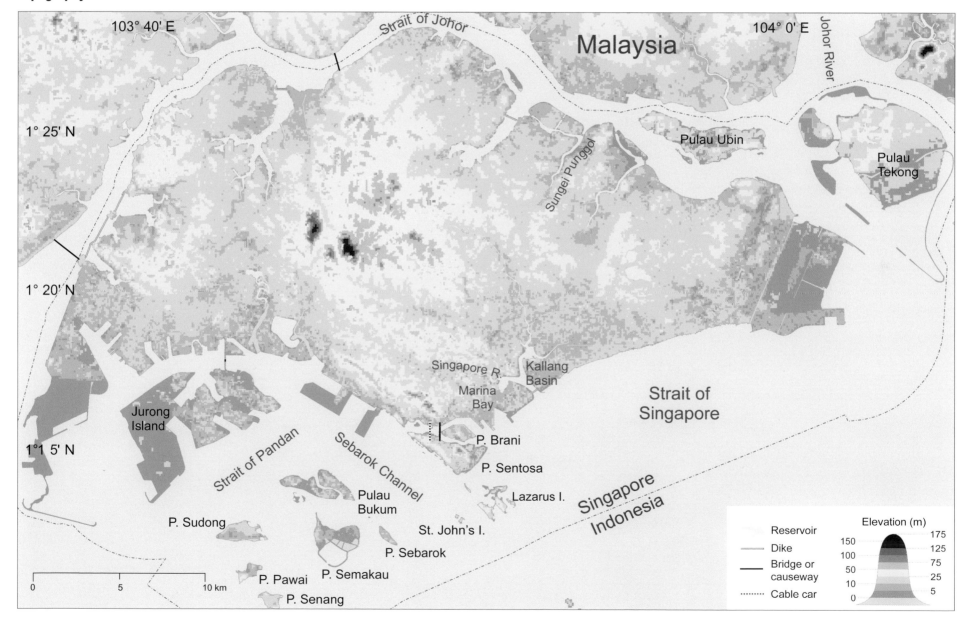

To facilitate the consultation of the series of plates that follow, we have grouped here what are essentially reference maps.

The first, dealing with topography, is a virtual one or rather an historical and composite one. The data was largely extracted from 1966 topographical maps and from a map produced by Wong Poh Poh in 1966 at the Geography Department of the then University of Singapore. Given the lack of comprehensive topographical maps currently available, we have adapted that "historical" information to fit the contemporary contours of the main island. The map shows that, while Singapore was never really mountainous, it was and still remains far from flat, with several ranges of hills structuring the territory, particularly at its geographical core. This renders even more astonishing the extensive transformations administered, so to speak, to the island over the last five decades (pp. 20 and 23).

The districts correspond to the so-called DGP (Development Guide Plan) zones defined in the Urban Redevelopment Authority (URA) Development Guide Plans. They are grouped under five regions, to which must be added the central area and two Water Catchment districts.

The administrative districts of Singapore. Generally divided into North, North-East, East, Central, and West region, with a "Central area" , referred to as "downtown" and also representing the main tourist/ historical district.

MAP ON P. 15: The topography of Singapore. Most of Singapore's hills had by the 1990s been levelled to provide land for its reclamation projects.

Administrative Districts

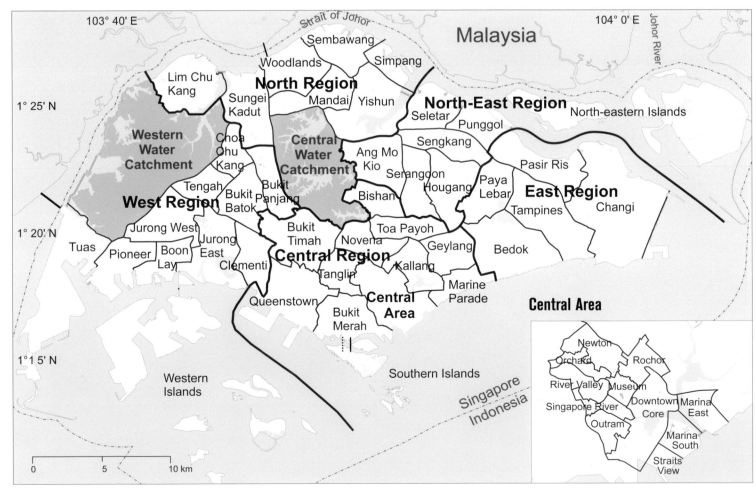

They were used in the 2000 census. This map therefore provides a number of place names, many of which will not be repeated in the rest of the book, and are therefore useful for the consultation of several of the maps that it contains.

The third map serves to illustrate the extent to which Singapore is highly urbanised, to a point where it does deserve the appellation "city-state". It represents a gross simplification of the information provided in the very detailed URA 2015 Land Use map (p. 130). Thus the "built-up" land use category encompasses nearly everything, except open and green spaces, agricultural areas, areas to be developed such as those recently reclaimed from the sea and, finally, military areas. This non-built-up land is rather extensive in districts such as Western Water Catchment (the largest), Mandai, Seletar, Paya Lebar and Changi (p. 124).

Finally, this plate can also serve, as do several others, as a reminder of how close the island of Singapore is to its less urbanised neighbours, Malaysia and Indonesia, and how sensitive maritime border issues are. This is particularly the case with the very narrow Johor Strait and the shipping lanes that literally surround Singapore.

A City-State

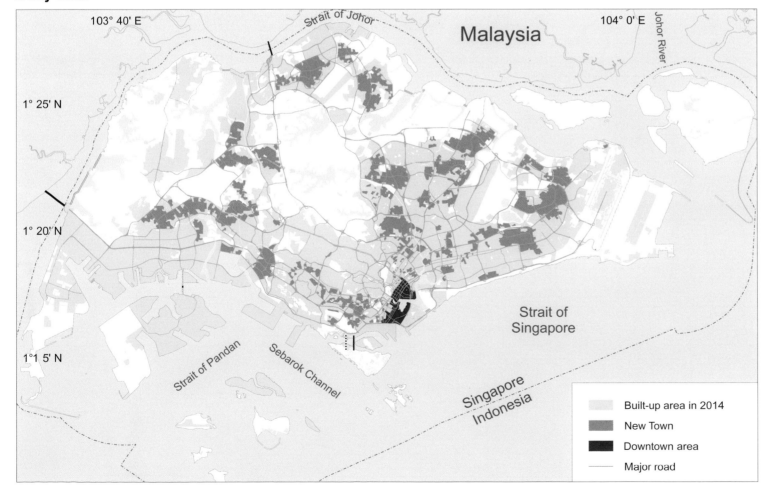

Main built-up area of the city state. Most of Singapore's limited land is used intensively.

Chapter 2
Taming Nature

"I sent research teams to visit botanical gardens, public parks
and arboreta in the tropical and subtropical zones (…).
Our botanists brought back 8,000 different varieties
and got some 2,000 to grow in Singapore. (…).
Greening is the most cost-effective project I ever launched."

– Lee Kuan Yew, 2000, pp. 177–8

Planned environmental transformations are not new to Singapore. Shortly after settling on the island in 1819, the British remodelled the banks of the Singapore river estuary, filling in the surrounding swamps. Throughout the nineteenth century, along with the harbour's development, came further drainage, land filling, and even land reclamation along the coast extending southwest of that estuary. The earth used for this seaward expansion was collected from the hills in the island's interior. Mostly confined to the harbour and city area, such earthworks continued during the first half of the twentieth century. Some transformations also took place on the north shore of the island, near the 1 km-long causeway leading to the Malay Peninsula that was constructed in 1923, as well as further east to accommodate the Sembawang air and naval base, which although only completed in 1938 had been planned in 1923.

But it is the postcolonial administration that initiated the real overhaul. Beginning in 1959, and gaining intensity after the founding of the Republic of Singapore in 1965, these environmental transformations were to become a permanent process. It first concerned the very shape and dimension of not only the island of Singapore itself, but also most of the 70-odd smaller islands that were part of the national territory. By then, size and shape were taken to task, as were regional and local functions. Some areas previously covered with swamps, such as most of the south-western portion of the island, were filled in to accommodate an industrial estate. On the eastern flank,

land was gained on the sea so as to allow for the relocation of what was soon to become one of Asia's key airports. Recently, also on the eastern flank, the two Pulau Tekong islands were united and extended in order to provide the Singapore Armed forces with a large training platform. Further out at sea, islands were evacuated and some extended or regrouped to host the petroleum industry. Increasingly, much of the soil and sand needed for these large territorial transformations has had to be purchased from neighbouring countries and even more distant ones.

Among the more elaborate land transformation endeavours have been those related to water supply and management. The number of reservoirs has been substantially increased, largely by the diking of small river estuaries and the desalination of the resulting ponds. Along with the literal creation of fresh water reservoirs, water catchment capacities have been enhanced throughout the entire island and several water treatment plants built.

Even the central urban area has been largely transformed to take in seawater bodies and integrate them in the city's landscape as well as in, after desalination, the freshwater supply network. More fundamentally, large portions of the island, in all directions, became susceptible to being transformed, with many designated to host New Towns as part of the very audacious housing policy that was to spearhead the national overhaul.

Permanent Secretary (National Development) Howe Yoon Chong visits Bedok reclamation site. Ministry of Information and the Arts Collection, courtesy of National Archives of Singapore.

6 Stretching the Land

"We have been changing Singapore's skyline. This project [East Coast reclamation] will change our shorelines and map."

– Lim Kin San, Minister for National Development, 2015

Cross section 1

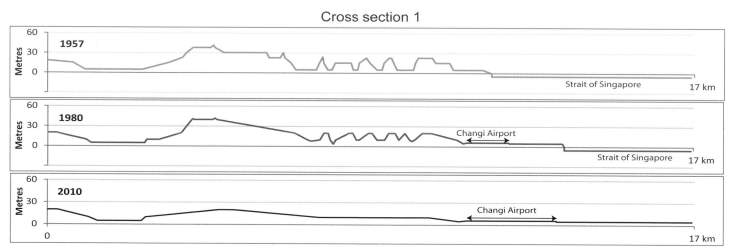

Cross section 2

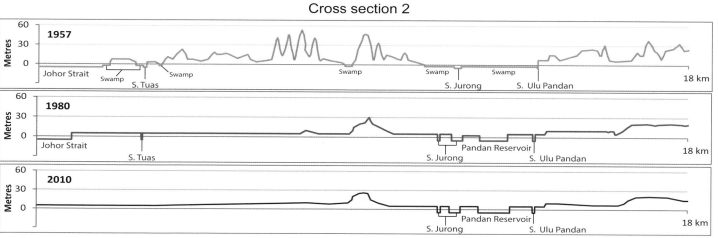

TOP: Two cross sections of Singapore and how its landscape has changed over 50 years: east and west.

Singapore Profile

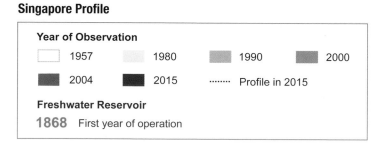

Year of Observation

	1957		1980		1990		2000

	2004		2015	....	Profile in 2015

Freshwater Reservoir

1868 First year of operation

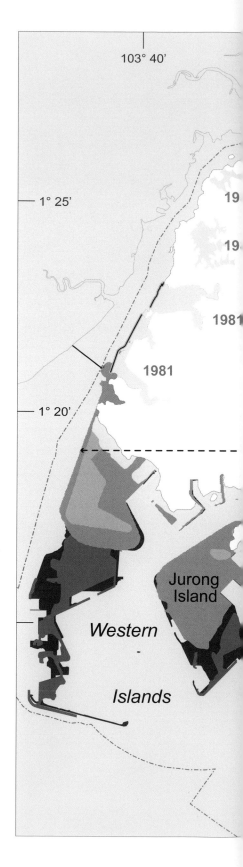

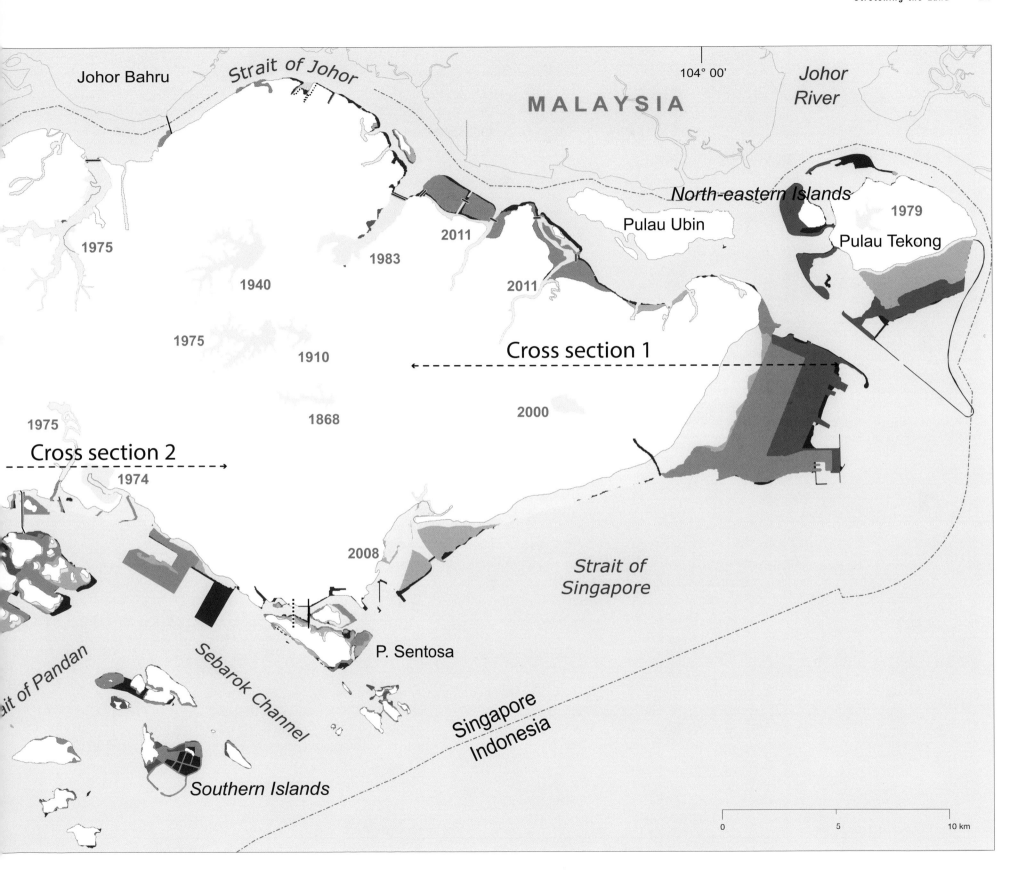

Johor Bahru

Strait of Johor

MALAYSIA

Johor River

104° 00'

North-eastern Islands

Pulau Ubin

1979

Pulau Tekong

1975

2011

1983

1940

2011

1975

Cross section 1

1910

2000

1868

1975

Cross section 2

1974

2008

Strait of Singapore

P. Sentosa

Strait of Pandan

Sebarok Channel

Singapore
Indonesia

Southern Islands

0   5   10 km

The Marina Bay Sands area, now a major tourist attraction and Singapore icon, is built entirely on reclaimed land.

Singapore's total land area from 1957–2015 (estimated).

Year	1957	1965	1980	1990	1999	2015
Land area in km2	581	586	618	633	659	719

In 1957, two years before it was to become a self-governing state, Singapore's land mass extended to over 581 square km. Between that year and 1965, when it became a fully independent nation, land expansion at the expense of territorial waters, which already had a long history, continued at a modest pace. But from 1965 onwards, expansion accelerated. Over the last 50 years or so, Singapore's land mass has gained nearly 135 square km at the expense of the sea. This represents a size increase of nearly 25 per cent.

Originally, and until the late 1960s, the main sources of landfill were local hills, which were levelled to provide the earth needed to "stretch" the island. But since then landfill has increasingly been obtained – purchased to be exact – from neighbouring countries (p. 23).

Land expansion has brought about a radical transformation of the relief and, even more, the outline of the main island and its satellites, particularly the Western Islands. On these are concentrated most of the country's petroleum tanks, refineries and other petrochemical infrastructure. The striking transformation of the island of Singapore's coastline is not only caused by its territorial extension through land filling or reclamation, but also by the closure of the estuaries of the main rivers draining the interior of the island. Also applied to the swamps encircling part of the island, particularly on the western flank, dike construction has also made possible the creation of additional fresh water reservoirs, such as those at Kranji and Seletar, once the newly created basins were desalinated (p. 26).

Expansion continues unabated today, with some of the more striking transformation being achieved on the east flank of the country, on the increasingly large island of Pulau Tekong (p. 24) as well as on the nearby Changi coastline with the massive expansion of Changi airport. It also concerns the laying out of the New Town of Punggol, where landscape transformations have literally reached new heights (p. 47).

7 Searching for Land

"Singapore has become one of the world's leading importers of 'stone, sand and gravel'."

– *UN Commodity Trade Statistics, 2014*

Having levelled large portions of its own landscape, and even if it has been continuously dredging the bottom of its territorial waters, Singapore has had to seek increasingly for land, or more precisely for soil and sand, from neighbouring countries and even distant ones. Among neighbouring ones, the first involved was Malaysia, apparently its sole foreign provider of sand until the early 2000s. Then, Indonesia was brought officially into the picture, with sand being purchased from the islands that make up the nearby Riau and Lingga archipelagos. While imports from Malaysia have continued over a few years, their quantity and relative importance have declined considerably, while other Southeast Asian countries have stepped in, with Cambodia and Vietnam as the leading providers well ahead of Burma and the Philippines.

Unsurprisingly, Singapore's hunger for landfill has created disputes with its nearest neighbours and come under heavy criticism from a number of circles. In 1997, Malaysia imposed a ban on sand exports to Singapore, although these have continued, both legally and illegally. In fact, the official imports of sand from Malaysia rose dramatically, particularly in 1999, 2000 and 2001. It seems that Singapore was then rushing to expand the Tuas peninsula southwards (p. 20), before negotiations with Indonesia over maritime boundaries got underway. In 2003, because of environmental concerns and, supposedly, border delineation issues, Indonesia banned the export of marine sand to Singapore; in January 2007, the ban was widened to include all types of sand and soil. It seems that since then the Indonesia–Singapore sand trade has been reduced to a trickle.

Malaysia has also protested against the expansion of the Tuas peninsula and of the island of Tekong, at both ends of the Strait of Johor. Nevertheless, expansion has continued even after the two countries came to a historical settlement in 2005. And the lucrative sand trade has also continued, but not from Malaysia anymore.

Nowadays, much of the criticism on the island republic's sand purchases concerns those made in Cambodia, where dredging is carried out mostly along environmentally fragile coastlines and river estuaries. According to the Ministry of National Development, Singapore stopped importing sand from Cambodia in November 2015 and is constantly on the lookout for new sources. Burma represents a new market – as well as a source of protests – and Bangladesh is also in its sights. Despite its size, Singapore is only behind countries as big as China, Germany and Turkey as an importer of "stone, sand and gravel", and is the world's leading importer of sand!

Main Sources of Natural Sand, 1989–2014

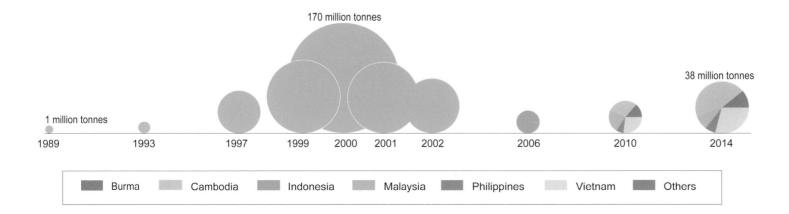

170 million tonnes

1 million tonnes

38 million tonnes

1989   1993   1997   1999   2000   2001   2002   2006   2010   2014

Burma   Cambodia   Indonesia   Malaysia   Philippines   Vietnam   Others

Singapore continues to import sand mostly from its Southeast Asian neighbours, though the volume and sources of sand have changed over the years.

8   Pulau Tekong as New Frontier

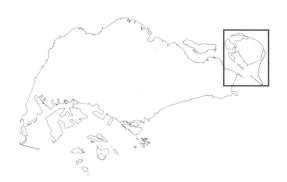

"It is necessary to build a second fortress or citadel in the Johor River estuary […] By constructing this citadel there (Pulau Tekong?), no enemy ships would be able to enter without being sunk, even though the Johor River is wide…."
— *Jacques de Coutre, c. 1600, in Borschberg, 2014, pp. 234–5*[1]

Whether at the heart of its "mainland", on its margins or its outlying islands, no portion of Singapore remains untouchable: any expanse of land can become a new frontier. Such is the case with the island of Tekong. As with agricultural frontiers elsewhere in the region, recent land pioneering in Pulau Tekong has involved relocation of pre-established inhabitants. Lastly, frontier development may raise objections from neighbouring countries, and such has been the case on the part of Malaysia over the expansion of Pulau Tekong.

By the late 1950s, Pulau Tekong, already the most extensive of Singapore's numerous outlying islands, was inhabited by some 4,000 people and was the site of large expanses of forests, swamplands and related flora and fauna. That approximate population figure was still valid in 1965, when the fully independent republic was founded. The island hosted communities that were predominantly Chinese (60 per cent) and Malay (40 per cent), the former primarily involved in farming, the latter in fishing activities. In 1958, seven Taoist temples and two mosques were located on Pulau Tekong, along with two Chinese cemeteries and eight Muslim ones (pp. 78 and 80), while by 1966 there were still on the island three primary schools, all Malay (p. 84). Population growth apparently continued, with sources mentioning a total of nearly 5,000 inhabitants by the late 1970s. By that time, the fate of the island began to change drastically as, through stages, it came into the sights of several land development agencies, such as the Housing and Development Board, the Ministry of National Development and the Urban Redevelopment Authority and, more recently, the Ministry of Defence.

As its inhabitants were gradually moved out and relocated on the mainland, the 25 square km island of Tekong became the object of a succession of typical planning and revised planning decisions, all of which involved massive land reclamation along its territorial margins. First, it was the Port of Singapore Authority – which has since been replaced by the Maritime and Port Authority of Singapore and PSA International Ltd – that undertook land reclamation on the island's southern shore during the late 1970s and early 1980s. Then, with the release of the 1991 Concept Plan, the government announced new plans for the development of Pulau Tekong. Along with the neighbouring Pulau Ubin, it would be the site of high-density housing and light industries development, all of which would need that these two islands be interlinked as well as linked to the mainland by the Mass Rapid Transit (MRT) network! This meant merging Pulau Ubin to Pulau Tekong Kechil and the latter to Pulau Tekong Besar.

While most of these extremely costly plans, running into billions of dollars, have since been abandoned, the merging of the small island of Tekong with the large one has proceeded but Tekong and Ubin have not been merged. The latter, a much smaller island, has been deserted by most of its permanent resident population and is increasingly earmarked for development as an outdoor recreation area, attracting mostly local tourists.

In 2002, Pulau Tekong's expansion by reclamation was contested by Malaysia, which submitted its protest to the International Tribunal for the Law of the Sea. The case went to court and by 2005 the two neighbours had come to an amicable agreement. Singaporean land reclamation along the shores of Pulau Tekong would continue with assurances given to Malaysia that it would have no effect on its own shores and territorial waters.

Since then, the island has continued to expand, to such a degree that, proportionately, it grows at a much faster pace than the mainland. It is now exclusively reserved for the training of the Singapore Armed Forces, as it will seemingly replace all other training grounds still located on the main island. Such exclusive military use for Singapore's largest outlying island, situated very close to Malaysia and at the western entrance of the Strait of Johor and near the mouth of the Johor River, is reminiscent of its pre-WWII history. Then under the control of the Changi Fire Command with troops being stationed on it, it was subsequently overrun by

the Japanese in February 1942. In fact, Pulau Tekong does represent a metaphor of Singapore's transformations as well as of its demographic, military, cultural and environmental history and heritage. This appears particularly striking with regards to its strategic and military function: the island is now a citadel, as foreseen more than four centuries ago by the Flemish traveller Jacques de Coutre.

Pulau Tekong is the largest outlying island in Singapore. It used to be a residential and farming island, with schools, places of worships, cemeteries and plantations. In 1987, the last of its residents were moved out. The island is home to the Basic Military Training Camp, where young Singaporean men train when they are doing their National Service.

---

[1]   The reference to Pulau Tekong, although highly probable, remains hypothetical. (Cf. Borschberg, 2014, pp. 234–5).

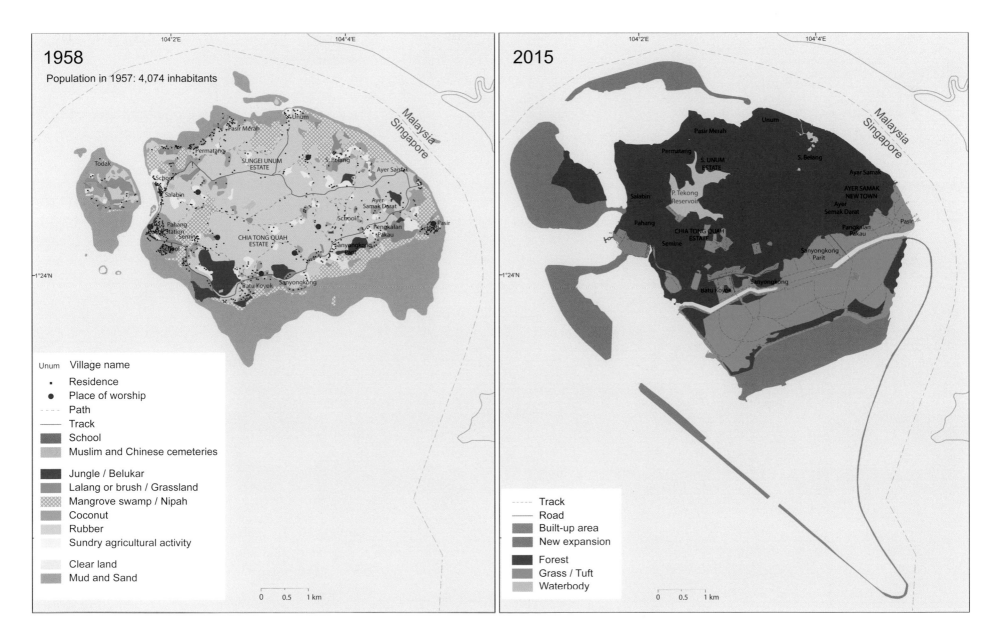

9 Collecting and Stocking Water

"As for water, we have alternatives […] we can manage."
– Lee Kuan Yew, 2000, p. 254

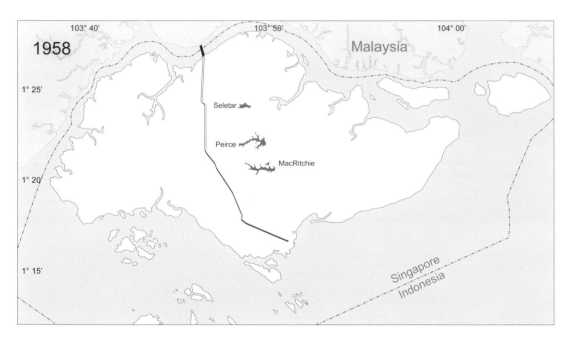

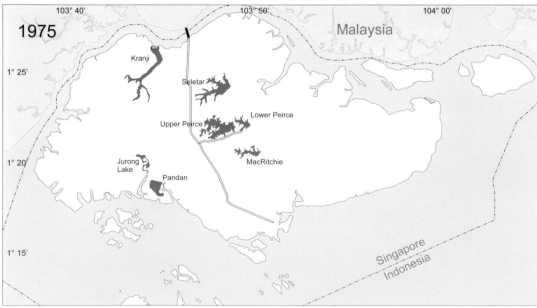

With its typically equatorial climate, Singapore's annual precipitation is abundant. The different regions of the island receive on average from 230 to 260 centimetres of rain, the drainage, redistribution and evacuation of which is insured by a constantly improving network of impressive canals.

Even with so much rainwater, the small island republic still falls short of self-sufficiency in fresh water supply. Despite the government's numerous campaigns calling on Singaporeans to save water, consumption increased much more rapidly than population. In fact, between 1960 and 2006, a period when the overall population (residents and non-residents) less than tripled, overall water consumption grew more than fivefold. The discrepancy was attributable to rapid industrialisation as well as to an overall improvement in the population's living standards. As a consequence, Singapore has had to acquire, at least until recently, about half of its water supply from the Malay Peninsula, an arrangement dating back to the colonial days. As early as 1931, a water pipeline was built across the causeway to bring in water from Johor. In 1961, even before the formation of the Federation of Malaysia (1963) – which included Singapore for two years – Singapore signed a water supply agreement with the state of Johor, valid in principle for 100 years. This agreement was modified the following year but still guaranteed a relatively cheap supply of water for Singapore. In the late 1980s, the government of the island republic actually purchased from Johor an entire 150 square km watershed, approximately 20 km inland, a third of which was transformed into a reservoir. The Singaporeans then built a water treatment plant on the spot, with 90 per cent of the treated water relayed by pipeline to the city-state and the rest resold to the Johor government, which apparently generated enough revenue to cover Singapore's costs.

Singapore's main reservoirs have been expanded since the 1950s, and new artificial water bodies created to collect and store rainwater for local use.

According to the Public Utilities Board, per capita domestic consumption has finally begun to decrease in recent years, but such has not been the case with consumption from other sectors. On the island itself, massive works have been launched to improve water storage facilities as well as water supply (p. 28). In 1958, the main island had only three large fresh water reservoirs, all located in its centre and covering less than four square km. Today, there exists 17 reservoirs – to which must be added several so-called "service reservoirs" – more dispersed and in some cases considerably larger, with a combined surface area of over 32 square km.

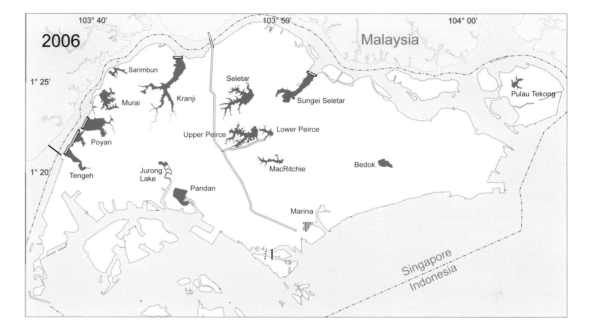

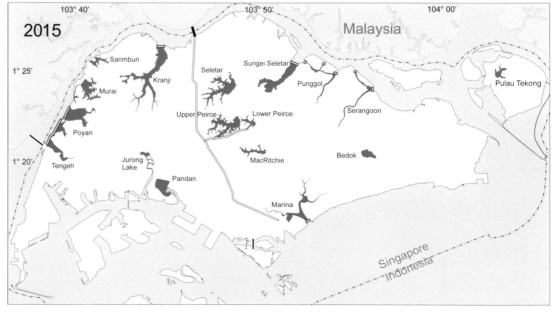

Reservoirs and Pipelines

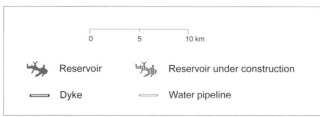

10 Diversifying Water Supply Sources

Over the last 50 years […] Singapore national water agency has built a robust and diversified supply of water known as the 'Four National Taps'.

Notwithstanding this impressive expansion of local water storage capacity (p. 26), including in the heart of the city (p. 34), the Singaporean water deficit problem remained very serious, taking added significance whenever attempts were made to renegotiate the previous water agreements with the Malaysian authorities. In this context, the government has continued to search for solutions and has been finding them! In the early 1990s, there was a plan to operate a water tanker service between the Indonesian island of Bintan and the Singaporean "mainland", but it apparently never materialised. Anyway, the city-state needed to find better and more reliable ways of achieving self-sufficiency. This quest has taken several forms, resulting in the Singapore's national water agency Public Utilities Board (PUB) developing a robust and diversified supply of water known as the "Four National Taps".

The *first tap* refers to rain water collected from local catchments, currently some 70% of the national territory and planned to grow up to 90% of it, the island being thus turned into a single huge catchment basin. In reality, this "national basin" is made up of several types of catchment areas: 1) a so-called Unprotected Catchment, located largely in the western part of the island and extending into the northern portion of the island's centre; 2) in the centre itself, a Protected Catchment, much less extensive, surrounding the three original reservoirs (p. 26); 3) and 4) east of these two Protected and Unprotected catchment areas, a very large urbanized catchment area itself divided into the Marina and Punggol-Serangoon basins; 5) a so-called Urban Storm Water Collection System, much less extensive and itself broken up into three interconnected

sections. It is expected that, with all its components, this very extensive first tap will provide some 20% of national needs.

The *second tap* brings in water imported from Johor, but considering Singapore's objective to reach self-sufficiency, it is likely to be turned down gradually over the years in favour of the three others.

The *third tap* supplies water coming from a number of Water Reclamation Plants (WRP). These provide, on the one hand Industrial Water, and on the other so-called NEWater, in fact used water that has been treated separately and is perfectly drinkable. By 2060 this third tap should fulfill up to 50% of all needs.

From the *fourth tap* comes desalinated water produced at the time of writing in two plants, both situated near Tuas on the southwestern shore of the island: it is expected to eventually fulfill up to 30% of national needs. And by then, Singapore would no longer depend on water supply from Johor.

Contrary to the popular statement that Singapore has no natural water resources, it has plenty, since the equatorial City-State receives an annual average of some 2.5 metres of rainfall. Of course this water needs to be channelled, stored, treated and properly used, but it is a resource. As with some of its other resources, such as its location, Singapore has been using its water resources with increasing competence. In fact, the island republic's research and development activities as well as concrete results in managing its water supply have become world renowned and have led, as with several other truly Singaporean breakthroughs, to the marketing of local experience, particularly during the Singapore International Water Week, held since 2008. In June 2015, it was attended by some 300 representatives from 35 countries.

Imported Water from Johor Malaysia 2014

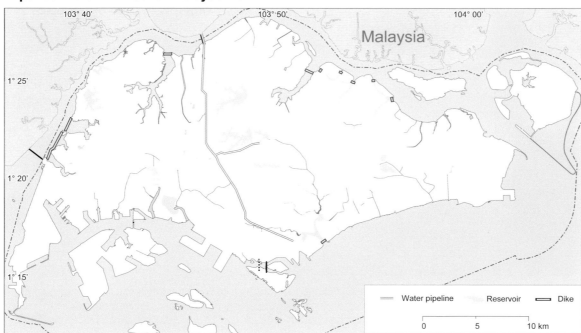

Singapore imports water from Johor by pipelines, covering 40% of its present needs (2014).

MacRitchie Reservoir is Singapore's first, and was completed in 1868. Photograph by Edwin Lee (Flickr user edwin11) shared on a CC-BY 2.0 license.

Reclaimed Water (NEWater) 2014

Treatment of reclaimed water will meet up to 50% of water demand by 2060.

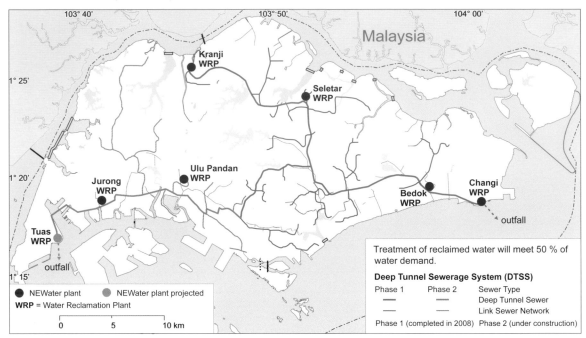

Desalinated Water, NEWater and Industrial Water Network 2014

Desalinated water from two plants can meet up to 30% of Singapore's current water demand.

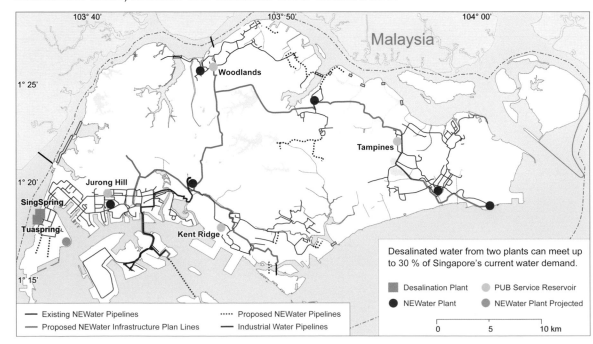

Catchment Areas 2014

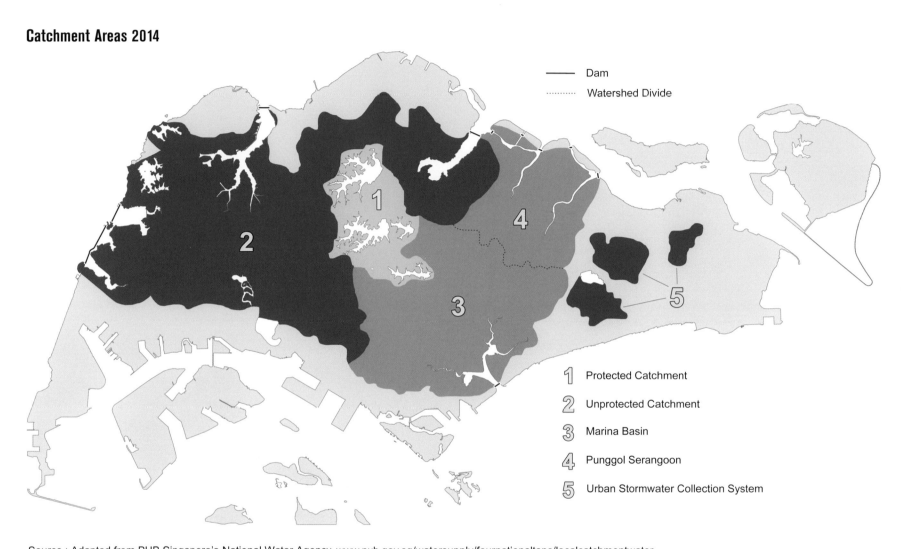

Dam

Watershed Divide

1 Protected Catchment

2 Unprotected Catchment

3 Marina Basin

4 Punggol Serangoon

5 Urban Stormwater Collection System

Source : Adapted from PUB Singapore's National Water Agency www.pub.gov.sg/watersupply/fournationaltaps/localcatchmentwater

11 The Garden City

"It was immensely better that we competed to be the greenest and cleanest in Asia."

– *Lee Kuan Yew, 2000, p. 177*

Although a vigorous tree planting campaign, covering not only the recreational areas but also residential, commercial and industrial ones, has accompanied urban expansion, natural forests have inexorably retreated. This occurred particularly during the 1970s, when mangrove forests were considerably reduced. In fact, stands of mangrove were essentially wiped out from the southern islands and all along the western and south-western coasts and estuaries, mostly to provide land for industrial installations. In the south-western part of the island, all inland forests also disappeared. The forest cover in the hilly centre of the island shrank during the same period, much of it giving way to reservoirs. But it is still substantial and retains today small strips of primary forest – in particular the 164 hectares Bukit Timah Nature Reserve, established in 1883, – while residual patches of mangrove are found in several locations along the northern shores. More substantial stands remain on the shores of Pulau Ubin and, especially, on the north-western coast of the main island, in the 130-hectare Sungei Buloh Wetland Reserve, which was given nature park status in 1989.

The retreat of natural forests and of agricultural land use has been balanced to some extent by a constant greening of the city and, in fact, of large portions of the island. Several of the remaining patches of natural forest have been either designated as or transformed into parks, with recreational facilities installed in some. Such is the case, for example, with the Bukit Timah Nature Reserve, located in the centre of the island and itself centered on the island's summit (Bukit Timah, literally Tin Hill, 164 m high).

Year	1960	1985	2006	2015
Area covered by natural forests in km²	37.8	28.6	22.64	~18.00

In addition, the greening of Singapore has benefitted from the establishment of a large number of new parks – currently numbering more than 300 and covering a total of nearly 20 square km – including some in the central urban area where remarkable ones had already been laid out during the colonial days. One of these is Fort Canning Park, at the top of Fort Canning Hill, a key historical landmark, called Government

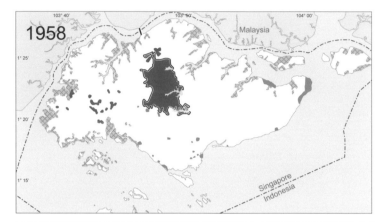

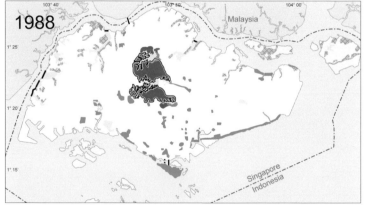

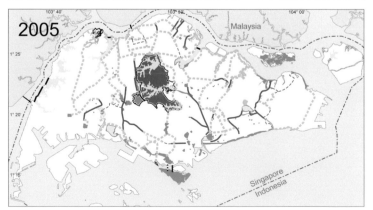

Hill in the early colonial days (p. 34), and Bukit Larangan (Forbidden Hill) prior to that. Another is the truly remarkable 74-hectare Botanic Gardens, which was first established in 1822 on the slopes of Government Hill and transferred to its present location in 1859. In July 2015, it was inscribed by UNESCO as a World Heritage Site.

The expansion of green parks has also occurred along the coasts, especially the southeast one, with the longitudinal East Coast Park. From 1971 to 1990 – at the initiative of then Prime Minister Lee Kuan Yew – an annual Tree Planting Day was held each November. After that, Prime Minister Goh Chok Tong launched the "Clean and Green Week" to expand the scope of Tree Planting Day. In this manner, hundreds of thousands of trees have been planted, including along roads and highways, to an extent probably unknown anywhere else in the world.

Finally, a network of pathways linking green spaces throughout the island is being put into place. These park connectors will eventually stretch to several hundred kilometres. The government obviously takes Singapore's green heritage very seriously. Since 1996, the National Parks Board has coordinated its maintenance and expansion. The latter exercises jurisdiction over all green spaces, whether the four nature reserves, parks and recreation areas or park connectors: in total, some 8,300 hectares of land (or 83 square km), nearly 12 per cent of the national territory.

Finally, to its policy of protecting and expanding its green tropical heritage, Singapore has added a modern touch. Over some 100 hectares of reclaimed land adjacent to Marina Bay, a complex of nature parks has been laid out by the National Parks Board. In addition to nature and recreation parks, Gardens by the Bay comprises a grove of giant metal trees as well as two huge glass and steel structures harbouring cooled nature conservatories. The Flower Dome and Cloud Forest Dome, along with the entire complex, have since their opening in the early 2010s become major attractions first for Singaporean residents and then for tourists.

The mangroves along Singapore's coastal areas have been removed over the years, and a complex network of park connectors and green corridors has sprung up. The next big park project is to convert the previous KTM railway track into a green corridor.

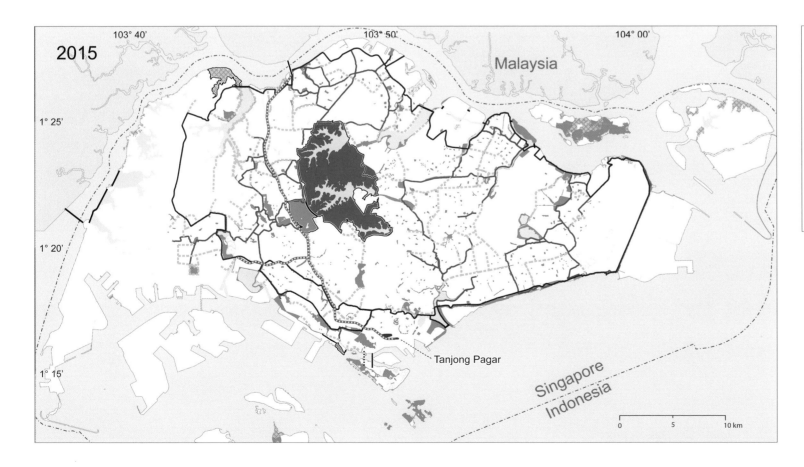

Mangrove
Older forest
Park
Nature reserve
Proposed park
Historic site
Completed park connector
Proposed park connector
Green Corridor
Round island route

12 The Sea in the City

Marina Barrage has become a showpiece of engineering and environmental innovations at the core of a network of walkways, parks and playgrounds.

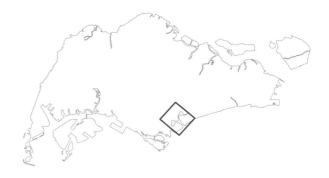

The seaward expansion of the Singaporean shoreline concerns not only the island's south-western and south-eastern reaches, which until the 1950s were still largely rural, but also the urban core, the Central Business District or Downtown Core. The latter has developed on either side of the mouth of the Singapore River, where the original colonial city and its Asian "suburbs" were laid out on a narrow coastal plain at the foot of low lying hills. Today the river, which until the early 1970s flowed directly into the sea, discharges its waters into a bay surrounded by two large expanses of reclaimed land, Marina East and Marina South. Its waters have been cleaned, and its course rectified. The same is true of the Kallang and Geylang Rivers, and the swampy and contaminated basin from which the Kallang River formerly emerged has been filled in.

Along with the Stamford and Rochor canals, all these waterways now converge on Marina Reservoir, the island's only truly urban reservoir. It was formed following the construction of the Marina Barrage across the Marina Channel, which flows between the expanses of reclaimed land called Marina East and Marina South. The barrage was officially opened in 2008 and within two years the entire reservoir's waters had been desalinated through natural replacement by rainwater and become an integral part of Singapore's network of fresh water reservoirs (p. 28). Along with the barrage this downtown reservoir plays a multifunctional role, as do so many of Singapore's land development ventures. Not only does Marina Reservoir boost the size of Singapore's fresh water supply at the very heart of the city, as proudly claimed by the Public Utilities Board, Marina Barrage represents a key component of a comprehensive flood control scheme to alleviate flooding in the low-lying areas in the city such as Chinatown, Boat Quay, Jalan Besar and Geylang. In addition, the 350-meter long barrage has become a showpiece of engineering and environmental innovations located at the core of a network of walkways, parks and playgrounds, all planted with trees, bushes and flowers.

The grid of the former colonial city, neatly represented on the famous Coleman map of 1837, has been maintained, but has lost its waterfront. Paradoxically, in its quest to become an Equatorial Garden City, Singapore, the great harbour city, is increasingly turning its back on the sea, at least in its historical core. Beach Road, for example, is now three kilometres from the sea. With the integration of sea space into Singapore's territory, coastal swamps and estuaries along the island's periphery have been transformed into freshwater reservoirs through dike construction and desalination (pp. 26 and 28).

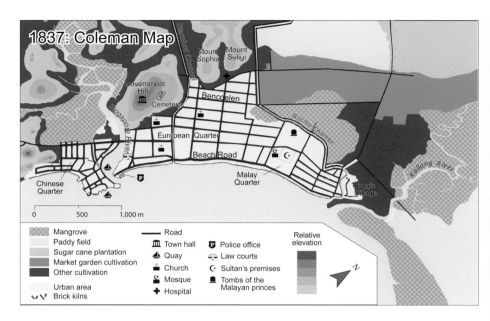

1837: Coleman Map

Mount Sophia
Mount Seligi
Government Hill
Cemetery
Bencoolen
European Quarter
Beach Road
Chinese Quarter
Malay Quarter
Rochor Channel
Kallang River
Bugis Village

0   500   1,000 m

Mangrove		Road	Police office	Relative elevation
Paddy field		Town hall	Law courts	
Sugar cane plantation		Quay	Sultan's premises	
Market garden cultivation		Church	Tombs of the Malayan princes	
Other cultivation		Mosque		
Urban area		Hospital		
Brick kilns				

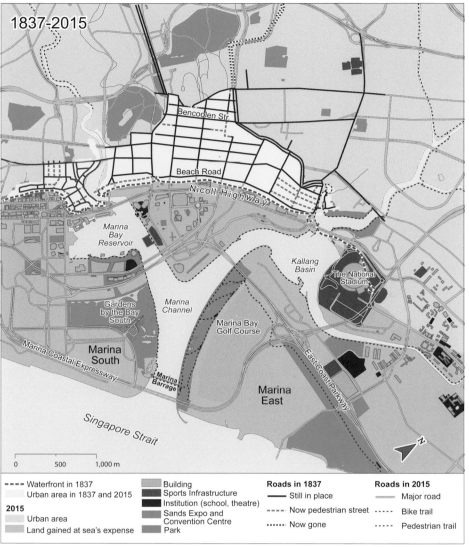

1837-2015

Bencoolen Str.
Beach Road
Nicoll Highway
Marina Bay Reservoir
Marina Channel
Kallang Basin
The National Stadium
Gardens by the Bay South
Marina Bay Golf Course
Marina South
Marina Coastal Expressway
Marina Barrage
East Coast Parkway
Marina East
Singapore Strait

0   500   1,000 m

---- Waterfront in 1837	Building	**Roads in 1837**
Urban area in 1837 and 2015	Sports Infrastructure	— Still in place
2015	Institution (school, theatre)	--- Now pedestrian street
Urban area	Sands Expo and Convention Centre	···· Now gone
Land gained at sea's expense	Park	

Roads in 2015
Major road
···· Bike trail
---- Pedestrian trail

Marina Bay is a 360 hectare extension to the Central Business District, and the extent of the land reclamation is obvious when 1837 and 2015 maps are seen side by side. It is designated as a mixed use area for commercial, residential, hotel and entertainment purposes.

Covering 101 hectares and opened in 2012, Gardens by the Bay is wholly built on reclaimed land and just beside Marina Reservoir.

Development of the Marina
Bay area, and its built-up space
between 2001 and 2014.

29/9/2014

Chapter 3
Repositioning People

"The Singapore of today is the direct result of the modernizing zeal of the first generation of post-colonial leaders. They are men not given to cosmic dreams or grand ideologies. They are practical realists. Thus the achievements of the state, the vindication of its policies and the symbols of success that now exist are in the form of the material development of the state, as for instance in the ever-present high-rise public-housing estates. Indeed public housing in Singapore is the single most visible index of the government's outstanding performance; it is the de facto monument to the People's Action Party government's success. There is no need for another."

– Tay Kheng Soon, 1989a, p. 860

Since the early 1960s and particularly after independence in 1965, the Singapore authorities have kept the national territory in a state of permanent upheaval. This has been rendered possible thanks to special assets, including a sizeable land bank inherited from the colonial administration in 1959 (crown land), equivalent to a third of the country's surface area. To this was added an additional 10 per cent of the country's territory, made available as the British military bases were gradually closed down (1968–71), a few years after Singapore separated from Malaysia (1965). Even more important for development or redevelopment, a series of laws – the Land Acquisition Ordinance (1955), Land Acquisition Act (1966), Property Tax Order (1967), and Control Premises Bill (1968) – gave the state legal powers to exercise eminent domain over practically the entire national territory. Armed with this authority, various government agencies – and in particular the Housing and Development Board (HDB) – were able to expropriate land as they deemed fit, and they exercised this power freely.

The result has simply been astonishing. All types of property in all parts of the island, rural as well as urban, were and remain subject to expropriation, always with due compensation. The relative intensity of operations does vary from one region to another, according to the different demolition and development programmes, with the urbanised centre of the main island frequently targeted. An expropriation site might be a vacant lot, an agricultural plot or a building, or an entire village or residential block whose occupants – in the earlier days they were often squatters – are invited to relocate while receiving financial compensation. Several sites have been the object of more than one operation of expropriation-evacuation-demolition-reconstruction. From 1965 to 1988, well over 1,200 sites were selected for expropriation and nearly 270,000 families were displaced, about a third of the country's population.

Since then, the number and the extent of transformation sites have probably been of the same order, particularly given the massive development of transport infrastructure networks, notably Mass Rapid Transit (MRT) stations. While the methods have evolved, and the costs and compensations greatly increased, the general principle still prevails. All over the national territory, dwellings may be acquired for upgrading or demolition and replacement, and when this occurs the occupants are required to sell and relocate. As people move, so do their schools, markets, places of worship and recreational areas. In this manner, some regions are quickly emptied of their population while others become heavily populated rapidly. And, in the process, one of the country's major social problem, that of ethnic concentration or even ghettos, is attended to. This has remained true ever since the great overhaul began in the late 1950s and even more since the 1965 independence.

Widening of existing canal in progress beside a row of attap-roof houses. Public Works Department (PWD) workers manually excavating earth from the base and side of the canal. Ministry of Information and the Arts Collection, courtesy of National Archives of Singapore.

13 Spreading out the Population

In 1957, before the great overhaul began, three quarters of the Colony's population, which then stood at about 1.4 million persons, lived within eight km of the mouth of the Singapore River.

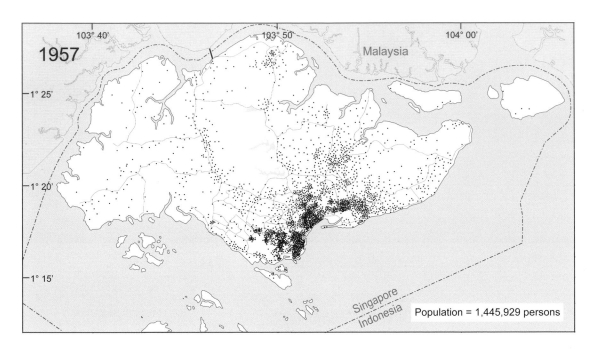

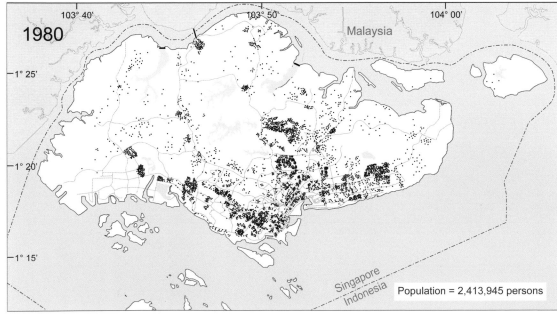

Starting in the 1960s, the relocation of displaced families has brought about a constant geographical redistribution of a national population whose growth has been held in check, although policies on that front have changed on many occasions. Comparing population distribution maps for 1957, 1980 and 2000 and 2010 (the most recent census year) provides a striking picture. [1]

In 1957, before the great overhaul began, three quarters of the colony's population, which then stood at about 1.4 million persons, lived within eight kilometres of the mouth of the Singapore River, concentrated in the urban core – notably in Chinatown – or in the nearby suburbs.

By 1980, the territorial distribution of the 2.4 million inhabitants appeared much less concentrated. The urban core and its immediate periphery had lost half of their residents, and the population there has since continued to decline. Conversely, around this zone a ring of population nuclei had taken form. These were New Towns, created by the Housing and Development Board, whose occupancy has since continued to increase steadily.

By 2000, when the resident population reached nearly 3.3 million persons, the overall occupation of the national territory appeared even more balanced. The decongestion of the old urban core had continued while several new outlying population nuclei had been created, notably in the North and North-East regions. Concurrently, several areas, including whole districts, had been largely emptied of their inhabitants.[2] Such was the case with the entire western flank of the island – particularly the three districts of Tuas, Western Water Catchment and Lim Chu Kang and – on the eastern side, the areas surrounding Changi airport.

Since then, as illustrated by the 2010 map, the extent and pace of population redistribution has been considerably reduced. Is Singapore finally approaching optimum territorial population redistribution? Nothing is less predictable, as New Towns are still expanding and additional peripheral areas are being devoted to exclusive military use (p. 24).

[1.]   The census figures used here take only citizens and permanent residents into account. In 2014, in addition to these, who numbered some 3,869,000, there were an estimated 1,599,000 non-permanent residents in Singapore, exactly twice as many as in the 2000 census year.

[2]   The method used does not allow for the representation of small numbers, one dot accounting for at least 500 persons. This means that the apparently empty regions can possibly be inhabited by small numbers of residents.

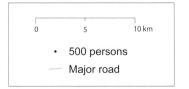

0   5   10 km

•   500 persons

⎯   Major road

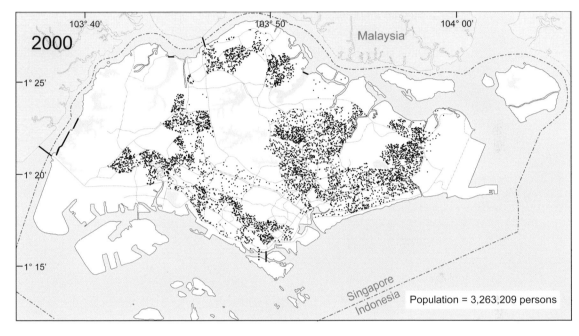

2000

Population = 3,263,209 persons

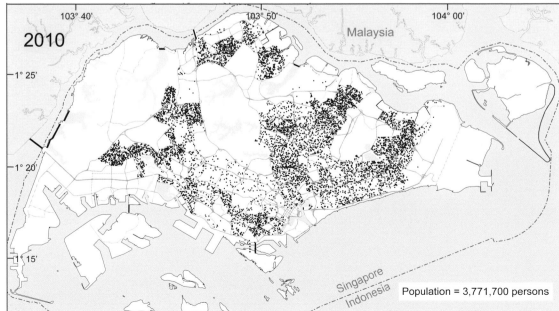

2010

Population = 3,771,700 persons

14 The Housing Question

"Another way to achieve this illusion of lower density is to maintain a sense of tidiness in layout, such that buildings are neatly aligned with one another."
— *Centre for Liveable Cities, 2013*

Public Housing Estates

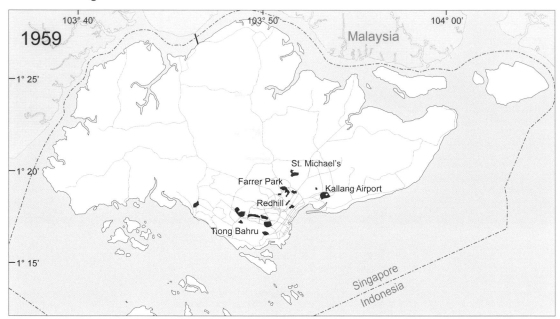

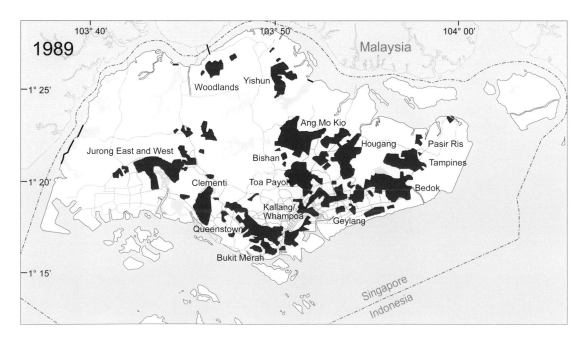

In 1960, when the Housing and Development Board (HDB) was established, the resolution of the "housing question" was placed on an equal footing with the restructuring of the economy and considered indispensable to national survival. Often praised, sometimes criticised, the creation of New Towns has since been the spearhead for the Singaporean territorial overhaul. A State within the State, the HDB has at its disposal an enormous budget along with wide powers to displace people and develop housing estates. Not only does it manage the largest national land bank, it is also by far the country's richest and most powerful landlord.

In 1959, when nearly a third of Singaporean residents were living in squatter settlements, less than 9 per cent of them were sheltered in public housing, still the responsibility of the Singapore Improvement Trust formed in 1927 under the colonial administration. By 1974, nearly 43 per cent of the population lived in HDB flats, and by 1989 the proportion had reached 87 per cent (i.e., 2.3 million persons). That percentage began to decrease as the share of private estates increased, along with the average Singaporean's purchasing power.

A total of 26 New Towns and Public Housing Estates are dispersed throughout the island, and cover more than 18,000 hectares, or a quarter of the national territory. Rarely is more than 50 per cent of that land actually devoted to residential use, the remaining land being designated for roads, commercial and educational use, green spaces, sports infrastructure, places of worship, etc. Population densities vary considerably from one New Town to another, depending on things such as the height and alignment of housing towers, and the size and design of recreation areas, but the overall result is usually a sense of density lower than the reality.

Since the steady transfer of the local population towards the public housing estates began, the HDB has introduced many changes within its empire. Types of infrastructure and

services offered are frequently modified and improved –
"upgraded" is the more common local expression – and so
are contractual agreements between the residents and the
Board. Since 1964, New Town dwellers have been allowed to
purchase their flats, and by 1988 nearly half of the apartments
constructed by the HDB were privately owned. In 2006,
among the 880,000 public apartments, the proportion of
those owned privately had reached 94 per cent. But this
ownership remains conditioned by the Board's prerogatives:
the latter still exercises full managerial autonomy as well as
a right of veto on any housing transaction. On its premises,
probably more than anywhere else in the island republic,
Singaporeans are closely supervised.

But the situation keeps evolving as private housing
estates are expanding more rapidly than HDB ones. Today,
New Towns, such as in Punggol, tend to juxtapose public
and private housing blocks (pp. 47 and 50). Overall, even
if home ownership ratio dropped slightly over the last few
years, it remained in 2014 at more than 90 per cent, one of
the highest in the world. That same year, about 81 per cent of
all Singaporean resident households still lived in HDB flats,
with some 90 per cent owning them. Some 13.5 per cent,
of resident households lived in private condominiums and
apartments, and nearly 6 percent on landed properties.

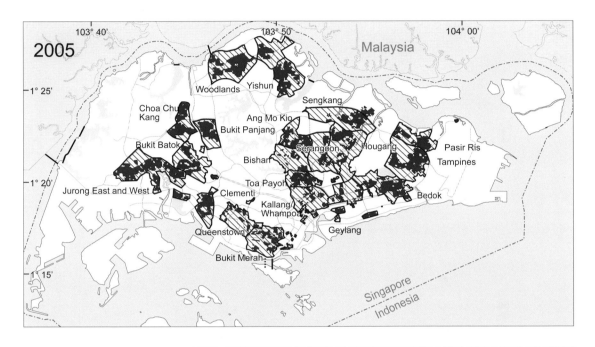

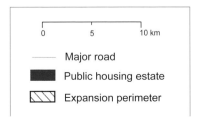

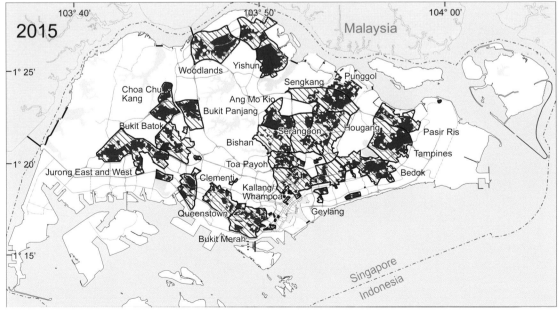

15 Housing Foreign Workers

Once again in the history of Singapore's economic expansion, massive relocation has come to the rescue of social management.

According to the 1970 census, taken a few years after the creation of the Republic of Singapore, foreign workers then accounted for 3.2 per cent of its workforce. Today, this proportion has reached nearly 40 per cent. What happened over the recent half century?

Firstly, a massive and rapid transformation of the landscape, whether "natural" or "institutional", has led to a call for huge amounts of labour.

Commercially Run Foreign Worker Dormitories

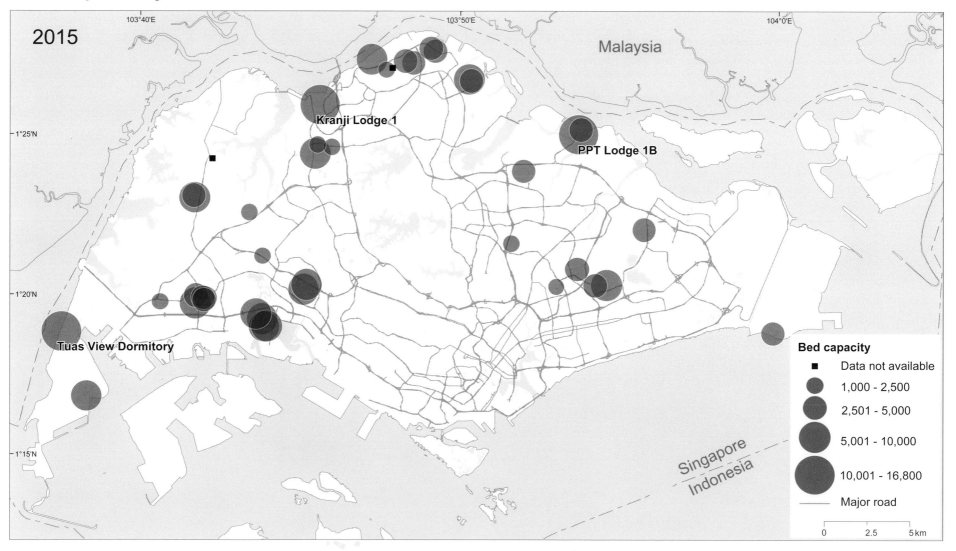

Secondly, the exceptional enrichment of a large number of Singapore citizens has meant that fewer are willing to take on certain jobs. Although the overall improvement of living standards has also increased inequalities among various categories or classes of citizens, as demand for labour kept growing, even the less favoured classes became increasingly reluctant to take on menial jobs, particularly in the construction – notably of the New Towns and private housing estates along with the MRT and LRT networks – basic services and maintenance sectors. Consequently these jobs have gradually been filled by foreign workers or employees. To this demand has been added that for domestic workers, almost exclusively feminine, an increasing number of households being able to afford hiring them.

Finally, the relentless demand for talent in several upmarket sectors, such as engineering and finances, also associated with the City-State's aspiration to become a global player, has led to the recruitment of so-called foreign talent. In fact, over recent years, for example between December 2009 and December 2014, the rate of growth in the employment of skilled workers and professionals (Employment and S Pass holders) has been much more rapid than that of construction and domestic workers (work-permit holders).

Detailed figures about the national origin of these non-resident workers are not made available by the Singaporean authorities, who prefer to remain discreet if not evasive. One thing is certain: unskilled workers come from a large number of countries including China, Thailand, Burma, the Philippines, Indonesia, India and Bangladesh. Construction workers are known to come predominantly from the Indian subcontinent, domestic workers from the Philippines, Indonesia and recently Burma, with highly skilled workers coming mostly from

A large dormitory specially built for foreign workers. Such dorms have as many as 16,000 beds, self-contained living quarters equipped as well with facilities such as indoor recreational rooms, TV rooms, indoor gymnasiums and outdoor game courts.

OPPOSITE: Most commercially run foreign worker dormitories are located at the outskirts of Singapore, where is it cheaper to house large numbers of workers.

Western countries, China and India. While the first group is by definition transient (i.e., ineligible to apply for permanent residence), among the others the possibility of applying for residence and even citizenship generally exists. But the rules and regulations are frequently modified, particularly in view of the increasing criticism coming from Singapore citizens who object to the competition on the labour market, particularly for higher skilled jobs.

Another issue that has become prominent over recent years is the housing of foreign workers, particularly the low-skilled ones. While domestic workers are by definition hosted by the households employing them, with construction workers the picture is much less clear, especially since public authorities, including the Ministry of Manpower (MOM) release very few figures. Firstly, an undetermined number of such workers rent a room, or even a whole flat to be shared among many, either in New Towns or even more in Little India and in neighbouring areas such as Geylang. This group does not always live under the best conditions. An additional and also officially unknown number are housed, legally as well as illegally, on work sites, particularly those located in peripheral areas of the island.

However, a fast increasing number of unskilled labourers, essentially those employed in the construction sector, rent a bed in private dormitories being built all over the island, mostly in peripheral areas, occasionally near or on industrial or public work sites. Forty-two of these dormitories have been identified and located. The number of beds they provide varies between a few tens to more than 10,000 in the largest ones; one of these, the Tuas View dormitory, offers 16,800. In these dormitory compounds, which are all registered with MOM, labourers are provided with breakfast and dinner and are generally able to bring their lunch boxes to the various worksites they are ferried to daily by fleets of buses.

The development of these dormitories – the largest of which are often found in isolated locations, far from major concentrations of Singaporean citizens and permanent residents, such as near the western shores of Singapore, on the Strait of Johor – was accelerated following the December 2013 workers' riot in Little India. Late 2015, it can be estimated that at least 150,000 workers (i.e., nearly two thirds of all work permit holders), reside in these nearly all-male proletarian towns. Once again in the history of Singapore's economic expansion, massive relocation has come to the rescue of social management.

BOTTOM: The foreign workforce (2010-2014).

The total number of foreign workers in Singapore and its proportion of total work force (1970-2014).

Foreign Workers, 1970–2014

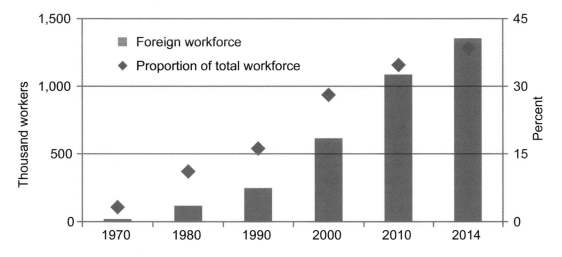

Foreign Workforce (Thousands)

Pass Type	Dec 2010	Dec 2011	Dec 2012	Dec 2013	Dec 2014
Employment Pass (EP)	143	175	174	175	179
S Pass	99	114	142	161	170
Work Permit (Total)	865	901	943	974	991
- Work Permit (Foreign Domestic Worker)	201	206	210	215	223
- Work Permit (Construction Worker)	248	264	293	319	323
Other Work Passes	6	8	9	11	15
Total Foreign Workforce	**1,113**	**1,198**	**1,268**	**1,322**	**1,356**

16 Laying out a New Town: Punggol

Singapore keeps reinventing itself, including in the housing realm, which remains at the centre of its territorial and social transformation policies.

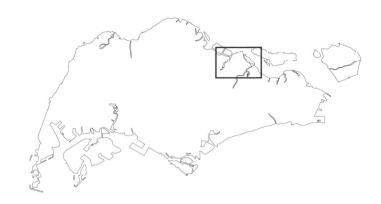

One of the goals pursued through the development of public housing estates was the decentralisation of residential, commercial and industrial infrastructure and activities. As the main instrument in the great Singaporean overhaul, the building of New Towns has inevitably been the object of elaborate planning. For example, the 1971 Concept Plan adopted by the Urban Redevelopment Authority (URA) – another powerful government agency – divided the main island into five planning regions (apart from the Central Area) and identified a centre for each one. One of the key principles underlying this new regionalisation was the creation of relatively autonomous activity areas, with places of work and residential housing brought closer to one another. The decentralisation goal remained important in the 1991 Concept Plan, with the further definition of sub-regional centres served by Mass Rapid Transit (MRT) stations. And as New Towns were created, older ones were constantly upgraded.

The actual deployment continues today, along with that of the MRT and Light Rail Transit (LRT) networks. A good example of the massive landscape overhaul entailed by the development of a New Town is provided by the case of Punggol. One of more recently developed New Towns, Punggol is situated in the northeast of the island, some 15 km as well as 15 MRT stations away from the Singapore River. It represents one of the island's peripheral localities with a fairly well-documented history, but also one that has been thoroughly transformed in less than over a relatively brief period of time.

Framed by the Punggol and Serangoon rivers, the Punggol peninsula had been home to Malay communities for several centuries. During the mid-1850s, Chinese settlers

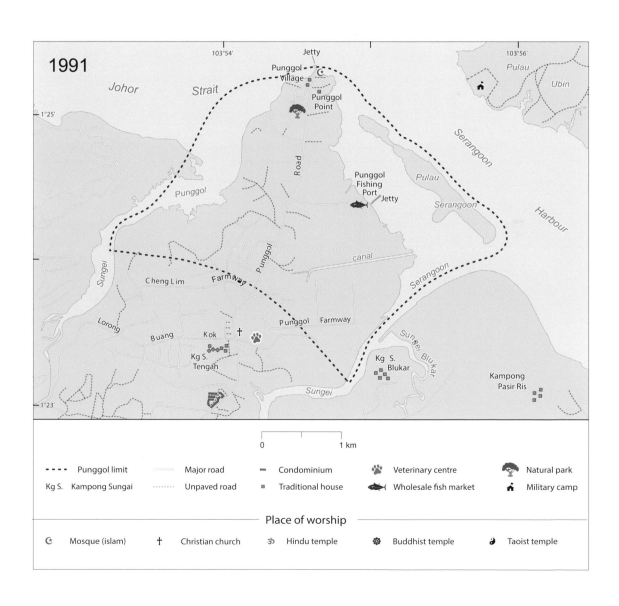

Punggol underwent rapid development after 2007, with many new built-to-order flats and residential facilities developed in the area.

moved in. In February 1942, some ten days after they overtook Singapore, the Japanese perpetrated a massacre of 300 to 400 Chinese Singaporeans on Punggol Beach, now identified by the National Heritage Board as a heritage site. Until late into the 20th century, the region remained nearly exclusively rural, most of its Chinese inhabitants – primarily Teochew – practising farming, generally poultry and pig rearing, with some of the few remaining Malay families still involved in fishing. At least until the early 1990s, a mosque stood at the end of Punggol Road overlooking the jetty on the Johor Strait. By then, the deagrarianisation of the area was well underway, with the dephasing of all forms of animal husbandry and the ephemeral attempt to encourage, instead, less polluting agricultural activities, such as hydroponic vegetable and orchid farms. But these did not survive long following urban expansion, which began with Sengkang New Town in the southern part of the peninsula.

In 1996, on the occasion of the August National Day rally, then Prime Minister Goh Chok Tong announced the development of Punggol New Town, calling it Punggol 21 and hailing it as a model for 21st-century towns! The town would contain a mix of private houses, executive condominiums and high-grade HDB flats grouped into smaller, distinctly designed estates. But the 1997 Asian Economic Crisis and the difficulties incurred by the local construction industry in 2003 delayed and slowed down the implementation of the grandiose plans. In 2007, under the name of Punggol 21+, they were revived by the Lee Hsien Loong administration. By 2015, the results appeared quite advanced and visible. The entire peninsula has been flattened and its drainage system redesigned. Dams have been built at both ends of Serangoon Island and at the mouth of Punggol River, thus allowing for the creation of two fresh water reservoirs (p. 26). And a huge New Town is now deployed, made up of HDB-run residential towers as well as privately run ones, well serviced by the expressway network as well as by the MRT and LRT ones (pp. 68 and 70). Singapore keeps reinventing itself, including in the housing realm, which remains at the centre of its territorial and social transformation policies.

Recreation areas, such as the newly opened Coney Island, as well as Singapore's newest university, the Singapore Institute of Technology, now cater to the large residential population.

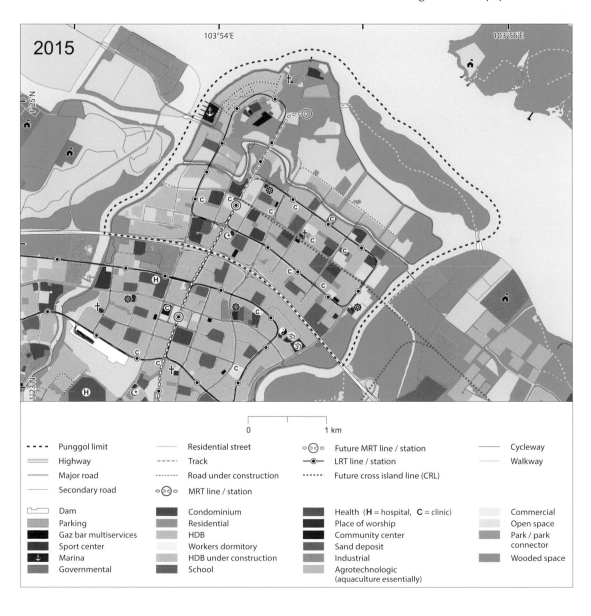

2015

▪▪▪▪ Punggol limit	—— Residential street	○⊙○ Future MRT line / station	—— Cycleway
≡≡≡ Highway	▪▪▪▪ Track	—⊙— LRT line / station	⋯⋯ Walkway
—— Major road	⋯⋯ Road under construction	⋯⋯ Future cross island line (CRL)	
—— Secondary road	○⊙○ MRT line / station		

Dam	Condominium	Health (**H** = hospital, **C** = clinic)	Commercial
Parking	Residential	Place of worship	Open space
Gaz bar multiservices	HDB	Community center	Park / park connector
Sport center	Workers dormitory	Sand deposit	
Marina	HDB under construction	Industrial	Wooded space
Governmental	School	Agrotechnologic (aquaculture essentially)	

17 Private Quarters

Private estates now house nearly 20 per cent of the country's permanent population. This is largely attributable to the growing affluence of Singaporean citizens and permanent residents.

The construction industry plays a crucial role in the Singapore economy. It is heavily involved not only in transport infrastructure construction but also in the development of the hotel industry and as a subcontractor to the HDB, private housing estates representing an additional lucrative market. Consequently, the sector has enjoyed a nearly uninterrupted boom since at least the mid-1960s, these private "quarters" now housing nearly 20 per cent of the country's permanent population. This is largely attributable to the growing affluence of a substantial proportion of Singaporean citizens and permanent residents. While in 1959, the number of private housing estates stood at 41, it had reached 846 by 1988, 2,071 by 2005, and 2,604 by 2015. Anyone residing in Singapore cannot help but be exposed to the over abundant publicity surrounding the constant development of private housing estates, or private condominiums. Besides "luxury" living conditions, the advertisements generally emphasise their locations near highways and MRT or LRT stations, along with swimming pools and green landscaping as part of the package.

Over the years, as the massive public housing estates (or New Towns) development programme was receiving all the limelight and a large chunk of public funds – as much as 30 per cent of development funds in the mid-1980s, a lot more than education and defence – the layout of small and increasingly private housing estates has continued unabated. These are made up of either more spacious and more luxurious residential towers, or only a handful of generally quite comfortable villas or bungalows. Disseminated throughout the urban network, often within close range of the New Towns, they have, at least until recently, been much less numerous in the outlying areas of the island, as middle-class Singaporeans tend to congregate within closer range of the historical urban core and the Central Business District.

However, over recent years, as the overall urbanisation process, including the development of transport infrastructure, has

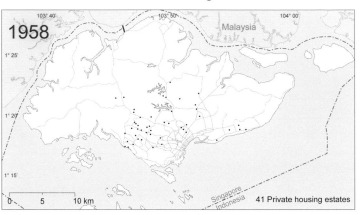

1958 — 41 Private housing estates

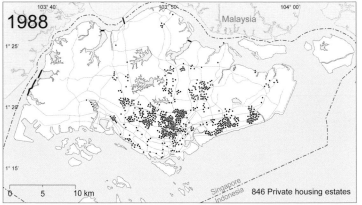

1988 — 846 Private housing estates

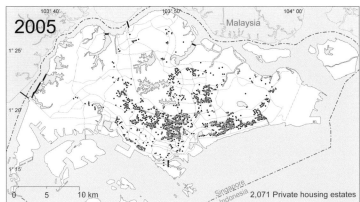

2005 — 2,071 Private housing estates

reached into the more outlying areas of the island, it has been accompanied by the creation of New Towns such as Yishun, Sengkang and Punggol. In the latter two cases, particularly that of Punggol, the location of private condominium towers, interspersed with HDB ones, constitutes an attempt to reduce residential segregation among various segments of the Singaporean middle class (pp. 40, 42 and 47).

There are several areas of Singapore that have public housing and private housing right next to each other. In fact, condominiums are sometimes almost indistinguishable from HDB flats. This is the Ghim Moh area in the west of Singapore.

- Private housing estate   —— Major road

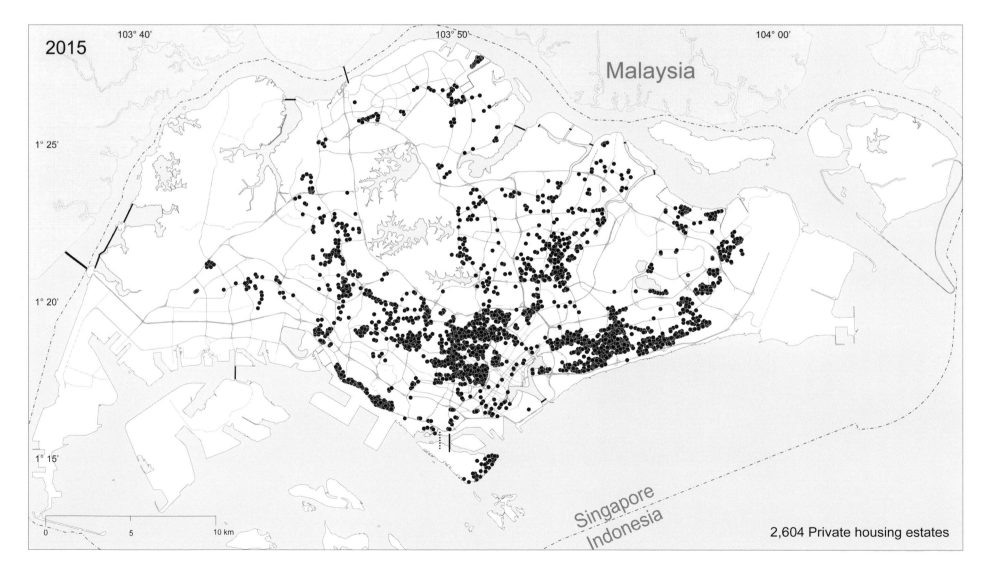

2,604 Private housing estates

18 Readjusting the Distribution of Ethnic Communities

The early colonial administration intentionally created ethnic concentrations, and others emerged as the entire island was gradually occupied.

OPPOSITE: The early colonial administaration created ethnic residential areas for the different races in the core city area, but by the 1980s the countryside also had similar ethnically dominated spaces.

The geographical distribution of ethnic communities has been a sensitive issue throughout Singaporean history. The early colonial administration intentionally created ethnic concentrations (p. 34), and others emerged naturally as the entire island was gradually occupied. In this way, ethnic areas that were almost ghettos developed not only in the urban core – for example with such well-known neighbourhoods as Chinatown, Little India and (Malay) Geylang – but also in the distant countryside, with the East Coast and the outer islands predominantly Malay, and Lim Chu Kang almost exclusively Chinese.

The creation of a sense of national identity transcending ethnic inclinations has long been one of the most commonly reiterated goals of the government. Hence, at least in part, the choice of English as the unifying language in the school system. Hence, also, a policy of population resettlement designed to break up ethnic ghettos, with quotas set to ensure minimal representation of each of the three main ethnic groups in each of the housing estates, taking into account national proportions. According to the 2010 census, these stood at nearly 74.1 per cent for the Chinese, 13.4 per cent for the Malays and 9.2 per cent for the Indians, whose proportional representation is the only one, among the three major groups, to have risen since the 1970 census, when it stood at 7.0 per cent. The precise pattern is hard to verify because the authorities do not reveal figures concerning detailed ethnic geographical distribution at the level of the administrative district, but it seems certain that no group has been exempted or favoured by the massive population redistribution of the last five decades, during which the great modern Singaporean overhaul has been implemented.

Moreover, reading and interpreting Singapore maps representing the redistribution of ethnic communities are complicated by the fact that the administrative grids changed between 1980 and 2000, and that by then many districts, more

than a dozen, were officially uninhabited, a circumstance that in itself illustrates the magnitude of the population displacement that has taken place. The task becomes even more complicated with the 2010 results as these are made available for a much larger number of census units, thus providing a much more detailed picture. Finally, given the wide discrepancy in the absolute numbers of Chinese, Malays and Indians, with the Chinese much more numerous than the Malays and the Indians, fair comparisons between the groups are difficult.

Nevertheless, while continuity seems to have been the norm, some changes are noticeable. The more striking ones revealed by the maps are as follows. First, the largest areas emptied of their resident populations were predominantly Chinese. Second, in the newly industrialised northern districts of Woodlands and Sembawang, Chinese predominance has increased over both the other groups. Third, major changes in the distribution and number of "places for the Malays", whether schools, cemeteries or mosques, are particularly perceptible because their original locations were often peripheral. For example, a rapid and drastic transformation occurred in the islands lying to the south of Singapore Island, which were evacuated by their inhabitants, most of them Malays. Fourth, in the central districts of Bukit Timah, Tanglin and Novena, the Malay presence has decreased in relative terms. Fifth, until 2000, the relative importance of Malays had however been maintained in the easternmost district of Tampines and Changi. Sixth, the latter district is also the only one where Indians had seen their relative representation increase slightly, possibly because of their greater presence in the service activities related to the Changi Airtropolis. Seventh, today, like both other groups Indians are no longer officially present as permanent residents in what could be called the greater Changi area. Lastly, the same can be said about the entire western flank of the island now devoted almost exclusively to industrial and military functions, as well as cemeteries and, very recently, foreign worker dormitories.

Chinese   1980   2000   2010

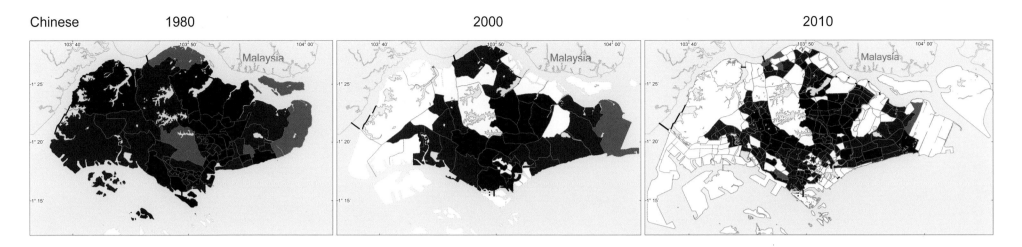

Malays

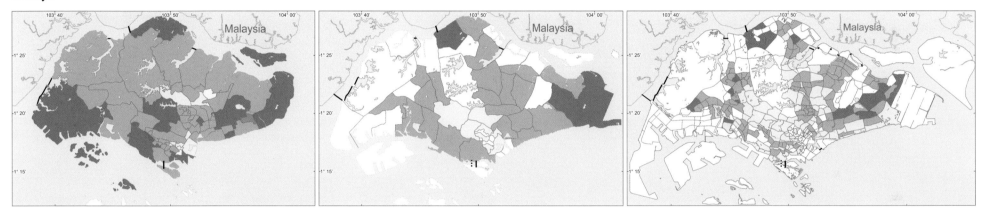

Indians

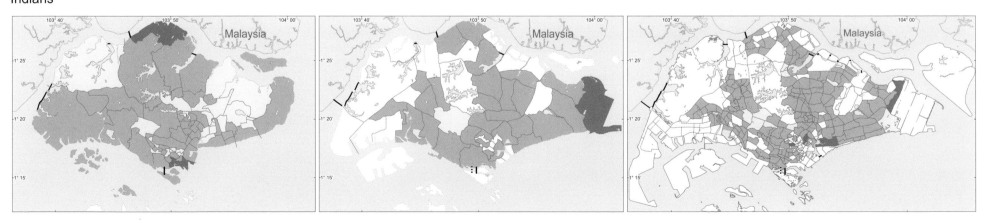

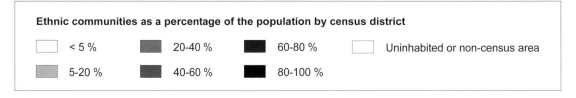

Ethnic communities as a percentage of the population by census district

| | < 5 % | | 20-40 % | | 60-80 % | | Uninhabited or non-census area |
| | 5-20 % | | 40-60 % | | 80-100 % | | |

Chapter 4
Reorganising Production and Circulation

These transformations result from and at the same time render possible a redefinition of the basic functions of production, be they agricultural, commercial or industrial, and circulation.

Since the early 1960s, the directions, shapes, length and nature of the coastlines, the topography, the rivers, water reservoirs – in fact the entire hydrographical networks along with the vegetation cover of the island – have been considerably modified. These transformations result from and at the same time render possible a redefinition of the basic functions of production – be they agricultural, commercial or industrial, and circulation – all of it closely tied to a new distribution of population, the latter tightly regulated. Concretely, these measures consist of the implementation of numerous urban renewal projects and the development of industrial and housing estates that in turn generate or call for, among other things, new transport networks and recreation infrastructure. This is achieved in a context where the local society and the national economy are fast increasing their links with the outside world. Illustrations are provided by the expansion of the harbour and airport installations as well as by the remarkable growth of the financial sector and even more of telecommunications (not illustrated here). As a result, nearly every locality is thoroughly transformed as it is earmarked for a new function, whether residential, industrial, commercial, recreational or even environmental.

By the early 1960s, land devoted to agriculture still accounted for nearly a quarter of the national territory. As such, it represented a crucial land bank at the disposal of Singapore planners who did not hesitate to tap into it. As agriculture was gradually phased out – or rather limited to very strict production conditions emphasising the reliance on agrotechnology – land was freed for other purposes such as industry, housing and transport infrastructure. The southwestern region of the island was earmarked as the ideal location for a huge industrial park that has since been constantly upgrading its activities through higher value production, service and research activities, with the more environmentally unfriendly petroleum industry largely confined to offshore islands. To power this rapid industrialisation, along with urbanisation and rise in living standards, it became indispensable to develop the energy sector. This refers to the process of transforming imported fuel into energy and conveying it throughout the island.

With an urbanisation policy relying on the construction of more than 20 New Towns located further and further away from the original urban core and implying a redistribution of population, the need to provide transport infrastructure for higher as well as lower income Singaporeans arose. The ensuing deployment of highway and rail networks – still ongoing and involving the construction of large Mass Rapid Transit (MRT) stations – continues to contribute to the highly visible overhaul of the island.

19 Rationing Agriculture

These highly productive farms mostly employ foreign labourers. They provide the city with a large variety of fresh leafy vegetables and some 30 per cent of its chicken eggs. Singapore's farm sector also supplies 15 per cent of the world market for cut orchids and nearly 20 per cent of the market for ornamental fish, for which it is the leading exporter.

In 1960, some 140 square km, nearly a quarter of the country's territory, were still devoted to agriculture and supplied nearly all of the city's requirements of poultry and pork, and half of its vegetables. By 1984, less than 50 square km of agricultural land remained. Two decades later, farmland occupied less than 7 square km, or about 1 per cent of the land, a proportion which has since been maintained.

Green Valley Farms, the first certified organic farm in Singapore, is located in Yishun. Photo credit to Jnzl (Flickr).

Agriculture's extremely rapid rollback was marked by two fundamental characteristics. The first was its total eviction from the immediate periphery of the urban core, the latter having for a long period remained essentially confined to the southern centre of the island. As was to be expected, expansion of the built-up area in other directions was initially achieved at the expense of agricultural land. The second was the partial regrouping, initiated during the 1960s, of agricultural activities in two specific regions of the island: the Lim Chu Kang District and, for a time, the Mandai hills of the North region. The contraction of agricultural land and the resettlement of farm operators were of course carried out in conjunction with urban renewal and industrial estates development.

Following the reduction in agricultural land use and employment, overall production declined considerably and irreversibly. Although needing little local space, pig rearing was completely phased out, primarily for ecological reasons, most notably water pollution. Two types of production, however – flower cultivation, particularly orchids, and some forms of fish rearing, particularly the offshore type – did however expand. But fundamentally, just like manufacturing more recently, agriculture and fishing were in many ways relocated abroad, with a growing number of Singapore entrepreneurs contracting out or investing in food production carried out in Malaysia and Indonesia but intended for consumption in the city-state.

More recently, agro-technology and agro-bio parks were established in Singapore itself, with surprisingly good results. Primarily located in the north-western districts of Choa Chu Kang and Lim Chu Kang, these highly productive farms mostly employ foreign labourers. They provide the city with a large variety of fresh leafy vegetables and nearly a third of its chicken eggs. Singapore's farm sector also supplies 15 per cent of the world market for cut orchids and nearly 20 per cent of the market for ornamental fish, for which it is the leading exporter.

The development of the agrotechnology parks by private contractors has been closely supervised by the Agri-Food and Veterinary Authority of Singapore (AVA). There are currently seven such parks, all located in the northern regions of the island, particularly in the northwest, the largest by far being the Lim Chu Kang Park, to which is adjoined a tiny agro-bio park. These parks cover a total area of nearly 700 hectares, nearly half of which is devoted to agricultural production on some 200 hundred farms.

Land Devoted to Agriculture

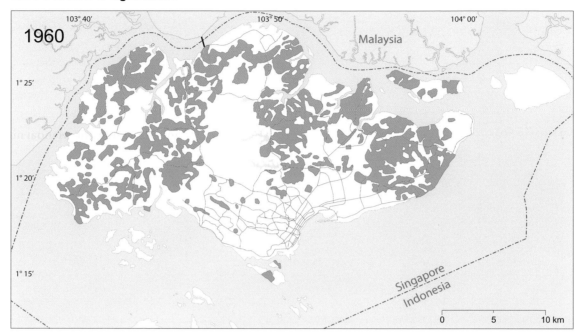

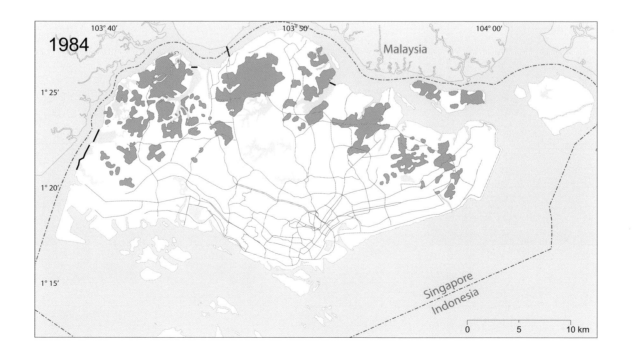

Sky Greens, located at Lim Chu Kang, is the world's first low carbon, hydraulic driven vertical farm. It uses green urban solutions to achieve production of safe, fresh vegetables, using minimal land, water and energy resources. Courtesy of Sky Urban Solutions.

Highly-Productive Agriculture, 1975–2012

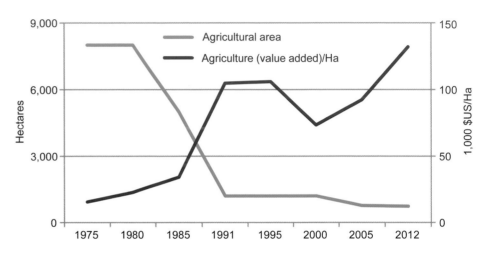

Farms in Agrotechnology Parks, 2001–2013

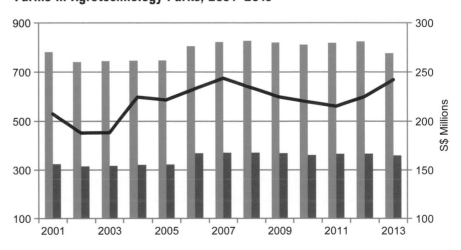

Extending over some 700 hectares, Singapore's six agrotechnology parks are located at Lim Chu Kang, Murai, Sungei Tengah, Nee Soon, Mandai, and Loyang.

Land Devoted to Agriculture

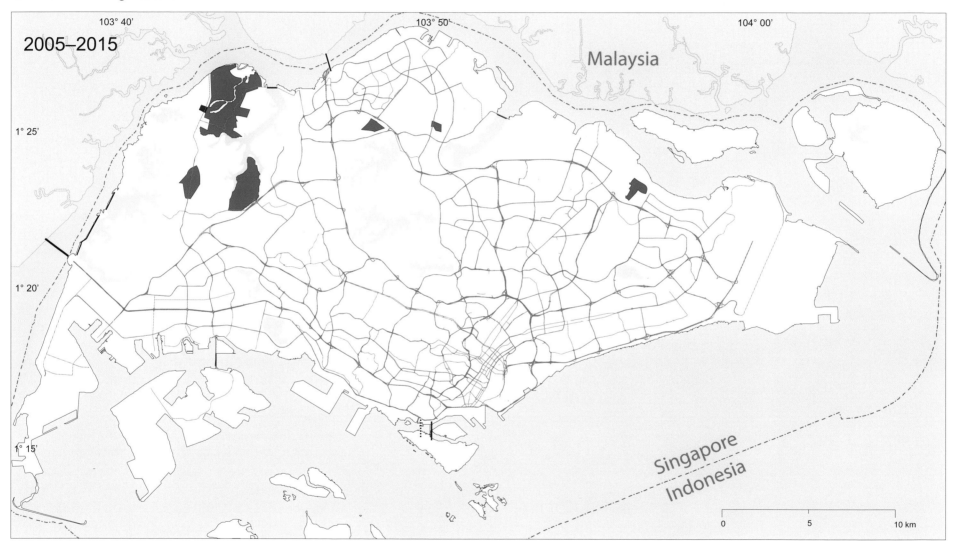

2005–2015

103° 40'   103° 50'   104° 00'

Malaysia

1° 25'

1° 20'

1° 15'

Singapore

Indonesia

0   5   10 km

20 Expanding and Consolidating Industry

"Our manufacturing sector has always been a strong anchor for our economy and a key source of our competitiveness and linkages to the rest of the world."

– *Report on the Committee of the Future Economy, 2017*

From the early 1960s onwards, Singapore's industrial production grew very rapidly. Between 1961 and 1989, its contribution to Gross Domestic Product (GDP) increased from 12 to 30 per cent. Over the same period, the GDP jumped from 2.1 to 55.3 billion Singapore dollars. Behind this expansion lay a vigorous industrialisation policy that granted large fiscal advantages to investors, local and foreign – multinationals in particular – in a context where local workers were offered intensive training and subjected to tight social control.

Singapore's remarkable industrial development could not, however, have been achieved without prior groundwork, literally. Before an industrial take-off could occur, room had to be made for free trade zones and industrial estates. Jurong Industrial Estate, the most important, was inaugurated in 1961, and by 1965, the year of Singapore's foundation as an independent Republic, 151 factories were in place, all of them profiting from pioneer status and tax rebates. At that time, the developed part of the estate covered 400 hectares (p. 62).

Since then industrial growth has never abated, although frequent adjustments and even industrial conversions have taken place throughout Singapore. These include the phasing out of the rubber industry, now flourishing where rubber is produced, in the Malay Peninsula, and the growth of the electronics industry. In 1968 the government created the Jurong Town Corporation (JTC) and gave it a mandate to monitor the development of industrial estates, then in full expansion and relying on the implementation of intensive land reclamation. By the late 1980s, the JTC was managing 24 industrial estates occupying a total of 76 square km throughout the island. These estates were home to more than 4,000 factories, half of them in Jurong alone. Among the other estates, the three largest ones, space wise – Kranji, Sungei Kadut and Woodlands – were located on the northern flank of the island.

But the most populous industrial estates, where flatted factories proliferated, were dispersed on the margins of the urban core itself: Ayer Rajah, Ang Mo Kio and, especially, Kallang Basin.

Since then, industrial activities have been partly relocated to neighbouring Indonesia and Malaysia (pp. 116 and 118), while in Singapore itself, industrial employment has levelled off and even decreased. The manufacturing front has been stabilised, with higher value-added industries favoured, and all sites located in peripheral areas. Jurong has retained its position as the pre-eminent industrial region, with a noticeable expansion of industrial land use in the entire southwest of the island, largely thanks to continuing land reclamation (pp. 20 and 62). The contribution of the manufacturing sector to GDP still stood at some 27 per cent in 2005 and 21 per cent in 2015. The electronics industry – which accounts for some 40 per cent of the global hard disc media production – remains the major contributor; in 2015 it accounted for more than a quarter of national industrial output and employed some 80,000 workers.

So, notwithstanding the continuing increase in the share of financial and business services to GDP, industrial production has remained central to Singapore's economy. It has received strong support from the Singapore government which has continuously and strongly increased its Research and Development budget, with, in the words of the Economic Development Board "the aim of making Singapore one of the most research intensive countries in the world", not only in the electronics sector but in several others such as the chemical and biomedical ones. Several of these research centres, such as Biopolis and Fusionopolis, are located near the National University of Singapore and several other institutions of higher learning, while others in the field of aerospace research and industrial development have been located near Seletar and Changi airport, both in the eastern part of the island. In these fields of endeavours, the territorial transformations are also widespread and intensive.

Finally, its experience and expertise with industrial development, and the fast-rising local salary scales have led Singapore to not only relocate much of its lower value industrial production to neighbouring countries (pp. 118 and 120), but also to market the said expertise worldwide, notably in China, India and the Middle East, where Jurong International is involved as a consultant in several industrial development projects.

OPPOSITE: Singapore's industrial estates started expanding in the 1960s to 1980s, with relocation to Malaysia and Indonesia accelerating in the 2000s.

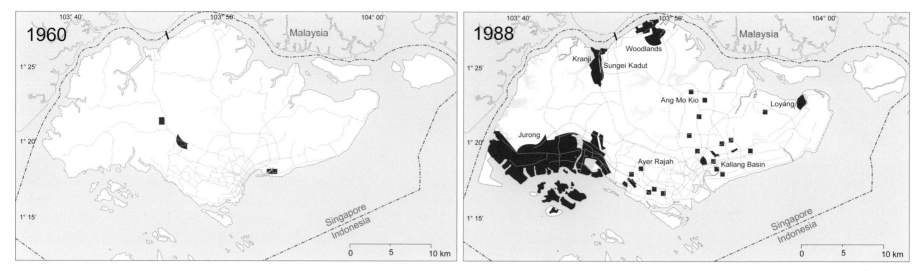

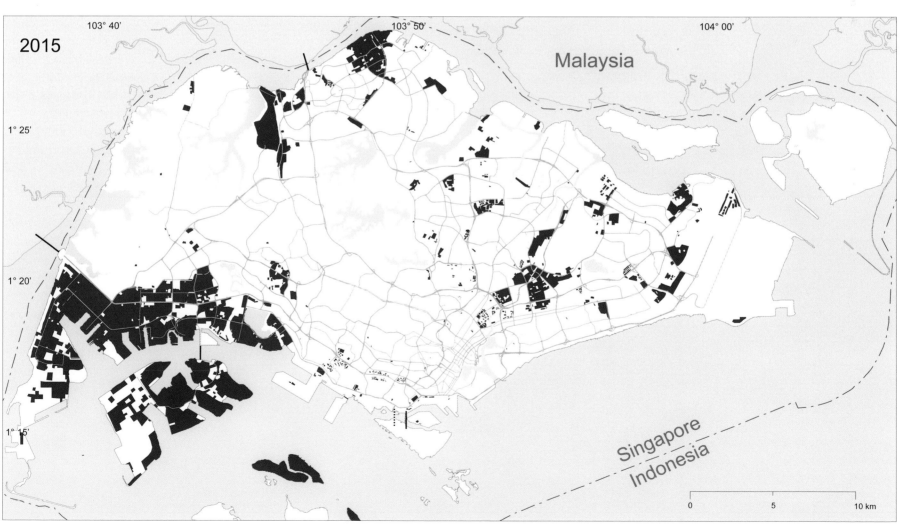

21 Jurong: From Mangrove to Industrial Estate to Urban Status

This will consolidate Jurong's status as an urban core, with its own skyline, increasingly competing with the traditional Central Business District.

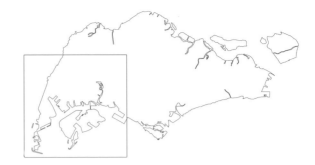

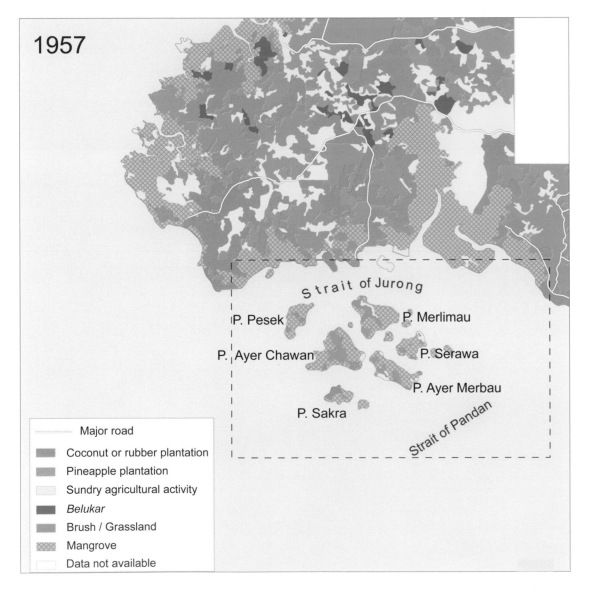

1957

Strait of Jurong

P. Pesek

P. Merlimau

P. Ayer Chawan

P. Serawa

P. Ayer Merbau

P. Sakra

Strait of Pandan

——	Major road
	Coconut or rubber plantation
	Pineapple plantation
	Sundry agricultural activity
	Belukar
	Brush / Grassland
	Mangrove
	Data not available

In the 1960s the overhaul of territory reached its greatest intensity in what was then the Jurong postal district, in the southwestern part of the island. The area had extensive mangrove forests, and low hills whose slopes were largely devoted to agriculture. By the early 1980s the hills – some of them 30 to 40 metres high – had nearly all been levelled and the extensive coastal mangrove areas thoroughly filled. Like a lone sentinel, a single hill remained. Today, its slopes are occupied by a park and its crest by a water tower and an observation platform. These structures overlook the industrial estate and the consolidated, straightened and ever expanding coastline.

Few areas of the island have been so completely transformed. As the core of the initial industrial effort, the district has seen the establishment of not only hundreds of factories but also a New Town, a residential estate created to house the workers. Its coast has become the new frontier for the expansion of the Singapore harbour, Jurong now possessing its own terminal and various forms of ship-berthing facilities. These now reach into the Jurong Islands, the area of which has nearly quadrupled as most have been merged into one very large and flat industrial platform, the Jurong Island. Until the late sixties, the waters around these

Named after the Jurong River,
the starting area was only
used for industry in the 1960s.

Land Use

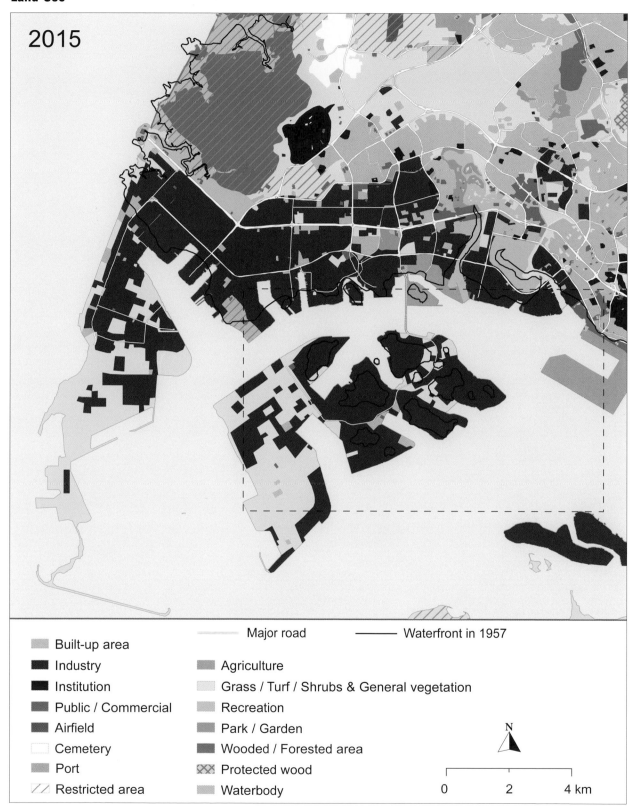

2015

▨ Built-up area	
■ Industry	▨ Agriculture
■ Institution	▨ Grass / Turf / Shrubs & General vegetation
■ Public / Commercial	▨ Recreation
■ Airfield	▨ Park / Garden
▨ Cemetery	■ Wooded / Forested area
▨ Port	▨ Protected wood
▨ Restricted area	▨ Waterbody

Major road ———   Waterfront in 1957 ———

N

0   2   4 km

Jurong has since evolved into an area that has both industrial and residential use. Two exciting forthcoming developments are the Jurong Lake Gardens, "Singapore's new national gardens in the heartlands", and the terminal of the Kuala Lumpur-Singapore railway line.

This 2012 nighttime view taken from Jurong Hill shows the petrochemical works at Jurong Island and the further development of Jurong as a logistics base. Photo by Allie Caulfield (Flickr user wm_archiv) shared on a CC-BY 2.0 license.

islands were still used for scuba diving. Now they share in the activities of the industrial estate and of the world's busiest harbour. The "mainland" itself continues to expand along the Tuas peninsula in the direction of both Malaysia and Indonesia but still well within Singaporean territorial waters. That industrial expansion is in fact closely associated with a parallel development on the Malaysian side of the Johor Straits, to which Singapore has been linked financially and structurally by a bridge opened in 2000, commonly called the "Second Link" (p. 120).

While until a decade ago or so, the name of Jurong remained closely associated with industry and harsh industrial landscapes, these have been to some extent softened with the systematic expansion of flatted factories. More fundamentally, particularly on the northern fringe of the industrial expanses, Jurong has become a full-fledged urban centre, a city in its own right. In the district of Jurong, public and private residential estates now cover nearly half as much ground as the industrial ones, while several new functions have blossomed. These include not only research labs but also recreational facilities and huge shopping malls and transport hubs, in particular for the MRT network with the junction of the East-West and North-South lines at Jurong East station (p. 70). Jurong's new centrality will be compounded by the slated opening in 2022, near this MRT station, of the railway terminus of the Kuala Lumpur–Singapore High Speed Rail. Jurong will consolidate its status as an urban core, with its own skyline, increasingly competing with the traditional Central Business District centered on the Singapore River and adjoining areas. And it illustrates a relatively recent tendency in Singapore's population redistribution policies: that of attempting to create peripheral growth areas, competing with the traditional central one.

22 Powering Singapore

Even before the 1965 Independence, the need to extend electricity supply
to the island's rural areas had become a major issue.

Providing Singapore with energy has long been recognised as a crucial task and has led to major initiatives, both to import and convert the fuel necessary to power the island's development. By the time of Singapore's independence, only two coal-fired power stations were in operation. The first, Saint-James, located in what was then Telok Blangah Bay, had been commissioned in 1926. The second was in 1952 in Pasir Panjang. In 1976, by which time other less centrally located power stations had been built – including in Jurong Industrial Estate and on the northern shore, near the causeway, with Senoko Station – Saint James station was decommissioned. Its original structure was

adapted for use as a warehouse for nearly 30 years, until 2006, when, it was transformed into one of Singapore's biggest entertainment complexes, housing a dozen bars, restaurants and nightclubs, located within walking distance of the HarbourFront MRT station.

Even before the 1965 Independence, the need to extend electricity supply to the island's rural areas had become a major issue. It took the Electricity Department ten years, from 1963 to 1973, to complete its Rural Electrification Programme. At the end of this term, electricity was available to everyone on the island. And as energy demand increased along with the affluence of Singapore's residents, but

Against a steady increase, Singapore's power consumption has moved from total reliance on oil to diversifying into natural gas and other energy sources.

Electricity: Sources and Consumption, 1971–2014

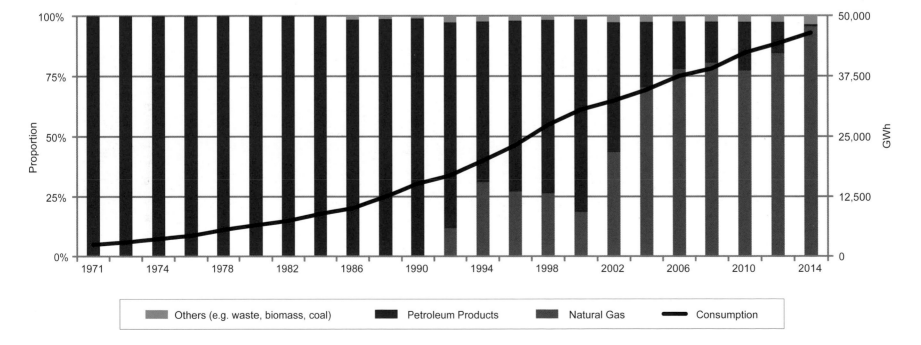

Legend: Others (e.g. waste, biomass, coal) · Petroleum Products · Natural Gas · Consumption

even more with the island's rapid industrialisation, the need to build power stations and to import the necessary fuel became a central preoccupation of the Singaporean authorities. This led to the construction of Pasir Panjang's power station B in 1965; of two stations in Jurong Industrial Estate in 1971 and 1974; of the Senoko station in 1976, on the northern shore, near the causeway; and in 1988 of the Pulau Seraya one, now part of the ever expanding Jurong Island, with power transmitted to the mainland through an undersea cable.

Since then, not only have some of these power plants been expanded, in particular in Pulau Seraya and Senoko, additional ones have also been erected. And the following features have emerged regarding the geography of power supply in Singapore. First, power plants have been largely removed from the urbanised areas including Jurong; on the mainland the only stations remaining are those found in the "Far West", in the Tuas peninsula, and in the "Far North" with the Senoko complex, by far the most important. Second, coal is hardly ever used anymore with natural gas by far the leading fuel utilised in these plants, some of which still have the capacity to burn petroleum. Third, Malaysia is by far the leading provider of natural gas, supplied by a pipeline which crosses the Strait of Johor. Fourth, a national power grid has been established, with a triangular structure broadly espousing the shape of Singapore's geography. Fifth, in 2001, an Energy Market Authority was established with the responsibility, among others, to regulate Singapore's electricity utility sector, but free market rules have increasingly prevailed with major energy providers consistently improving their competiveness and their own fuel consumption efficiency, notably with an increasing reliance on cogeneration plants (producing both heat and electricity). Sixth, so have most users, including the leading one, the industrial sector which in 2014 consumed over 45 per cent of all energy produced in the country, versus only 15 per cent by Singaporean households. Seventh, four incineration plants have been built over recent years. By burning waste, they generate electricity, while the ash remaining from the combustion is used primarily as landfill on Pulau Semakau, situated 8 km due south of the mainland, just beyond Pulau Bukum. Semakau Island is now Singapore's only landfill and is expected to be operational until 2035.

Notwithstanding Singapore's continued near total reliance on fossil fuels, the government has increasingly encouraged and subsidised research on solar power and other green energy sources. Initiatives that are particularly encouraged are those that deal with the use of solar power panels on rooftops, as well as architectural designs providing for better ventilation of these rooftops and vegetation covering of buildings having a cooling effect.

LEFT: St James Power Station was built in 1926 and was Singapore's first coal-fired power plant. It was shut down in 1962, and reopened as a nightlight and entertainment venue in 2006.

RIGHT: The Senoko Power Station, located in Sembawang, was commissioned in 1976 and is the largest in Singapore.

Singapore's Power Grid

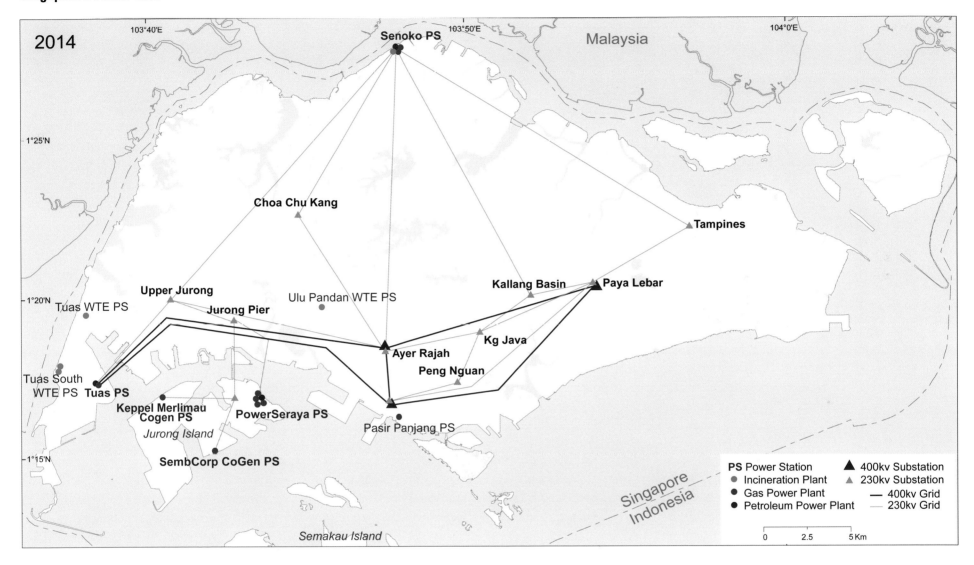

2014

Senoko PS
Malaysia

Choa Chu Kang

Tampines

Kallang Basin · Paya Lebar

Upper Jurong
Ulu Pandan WTE PS
Jurong Pier

Tuas WTE PS
Kg Java
Ayer Rajah
Peng Nguan

Tuas South
WTE PS · Tuas PS

Keppel Merlimau
Cogen PS
PowerSeraya PS
Pasir Panjang PS

Jurong Island

SembCorp CoGen PS

Semakau Island

Singapore
Indonesia

PS Power Station
• Incineration Plant
• Gas Power Plant
• Petroleum Power Plant

▲ 400kv Substation
▲ 230kv Substation
— 400kv Grid
— 230kv Grid

0   2.5   5 Km

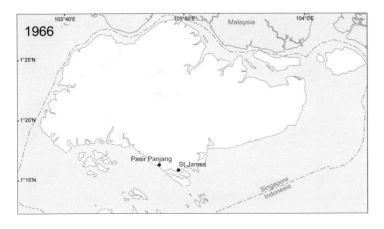

1966

Malaysia

Pasir Panjang   St James

Singapore
Indonesia

The number of power stations in Singapore has grown exponentially since the 1960s. Most of them however, are still located on the southwestern portion of the island, which remains its industrial core.

23 Petroleum Islands

At the core of the multifaceted petroleum and gas industry, stands the amalgamated Jurong Island, a typical example and possibly the most spectacular one of Singapore's capacity to overhaul endlessly its territorial components.

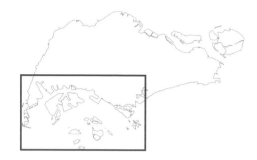

Starting in the 1960s, intensive construction of various types of industrial infrastructures occurred throughout the main island, to the point where they now occupy more than one tenth of its surface area. However, at least one sector achieved its own spectacular growth "extra muros", so to speak: the petroleum industry.

Initially needed for the rapid development of a city-state totally lacking in raw materials and heavily reliant on manufacturing, the energy industry has become one of the key sectors in the Singaporean export economy. Several power plants, including cogeneration ones, storage tanks, oil refineries and petrochemical installations have been stationed offshore, literally. Singapore's huge petroleum refining facilities are primarily located on the newly assembled island of Jurong – which has never stopped expanding through land reclamation – at a relatively safe distance from the urban core and the major concentrations of population and yet linked to the mainland by a causeway since 1999.

At the end of the 1950s, several of the western and southern islands were inhabited by rural communities, predominantly Malay and earning their living at least in part from fishing. All residents have since been evacuated, and their houses, mosques, temples and cemeteries relocated to the main island (pp. 78, 80 and 84). In their place, on islands often flattened and enlarged, or even fully constructed around coral reefs, were erected towering fuel tanks, refineries and other industrial structures. These are now nearly all regrouped on the sole island of Jurong, an impressive "floating" platform on which have also recently been dug underground rock caverns for storage of crude oil, now known as the JRCs (Jurong Rock Caverns)!

The powerful petroleum industry at one time accounted for as much as 40 per cent of the country's industrial production, but after peaking in 1982 this proportion has declined. Nevertheless, the petroleum sector's share of economic activity remains substantial. Perhaps more significant is the fact that commercial, technological and financial networks and flows linked to it have become crucial to the national economy. Singapore is not only a major refining and petrochemical centre but also the leading base in the region for offshore petroleum exploration and production. It is one of the top three export refining centres in the world, as well as Asia's leading oil trading hub. To this must be added its role as the leading bulk liquids logistics hub in Asia and, finally, as the world's busiest marine bunkering station. At the core of the multifaceted petroleum and gas industry, stands the amalgamated Jurong Island, a typical example and possibly the most spectacular one of Singapore's capacity to overhaul endlessly its territorial components.

The Southern Islands

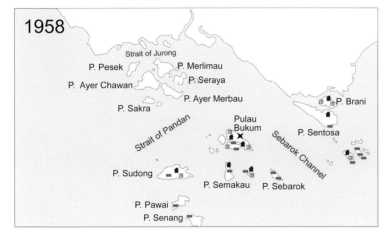

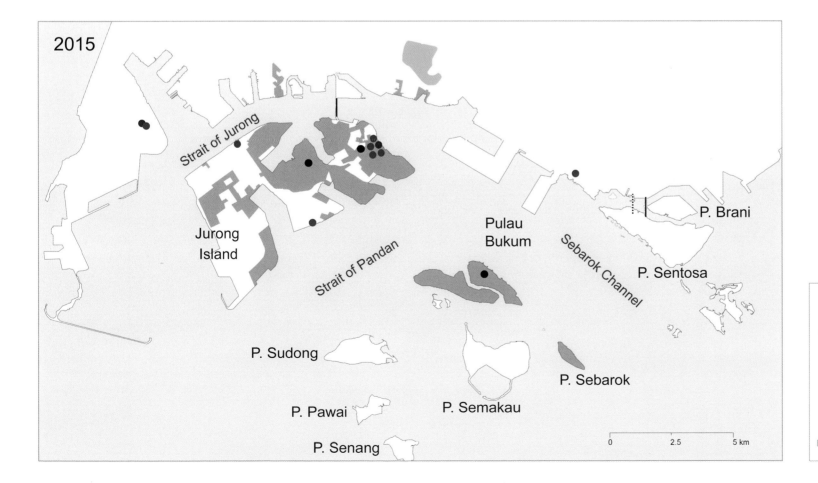

Initially used as places of residence, the islands of the west coast of Singapore have undergone major changes. Over recent decades, their land area has expanded substantially through land reclamation and the merging of the smaller islands into one big Jurong Island.

School
Mosque
Muslim cemetery

× Oil depot
● Refinery
● Petroleum power station
● Gas power station
Petrochemical infrastructure

24 Making Way for Cars

"In this manner, the Singapore government keeps pursuing two apparently contradictory goals. It keeps expanding a high quality road network, thus encouraging private car ownership among the more affluent classes. Concurrently, it invests even more in the provision of efficient public transport services. [Singapore] needs to decide how much further it wants to go on allocating priority to cars over people."

— Bruno Wildermuth, 2015, p. 20

The spatial expansion of the industrial estates and residential areas, particularly New Towns, along with the growing purchasing power of Singaporeans were accompanied by a steady rise in the number of motor vehicles on the island and a phenomenal improvement of its transport infrastructure. The constant expansion of the road network and particularly of expressways represents one of the most noticeable forms of territorial restructuring – some might say deconstructing – the island has undergone.

The Bukit Timah Expressway (BKE), a six-lane dual carriageway completed in the 1980s, east of the still operational Bukit Timah Road, provides a good example. The continuous increase in the number of cars circulating on the island's road network – to which some 12 per cent of the nation's territory is devoted – is occurring in the face of a series of disincentives decreed by the government. These include some of the highest car prices in the world (all cars are imported); very high taxes for ownership itself, including an expensive Certificate of Entitlement (or COE); high costs for fuel and parking; repeated increases in the levies collected for access, during peak hours, to the downtown core and to a number of expressways; and strict enforcement of traffic rules.

But the resulting paradox is apparent. For both the very entrepreneurially minded government as well as Singapore citizens, particularly those belonging to the middle and upper classes, seem to cope with this evolving situation. Notwithstanding the increasing congestion of the once famously smooth City-State traffic, Singaporeans continue to buy cars and to bid for the COE, while the government continues to contract out numerous and lucrative road construction projects. In this manner, it keeps pursuing two apparently contradictory goals. It keeps expanding a high-quality road network, thus encouraging private car ownership among the more affluent classes. Concurrently, it invests at least as much in the provision of public transport services. Indeed, for those who cannot afford car ownership and use, there is the equally expanding and improving public transport system and a large fleet of taxis providing a generally efficient and relatively cheap service. But, and this appears as a real paradox, much less is provided in terms of infrastructure for pedestrians.

OPPOSITE: The first expressway in Singapore was the Pan-Island Expressway; its construction began in 1962. By 2015, there were 11 expressways spanning the island, with the newest one being the Marina Coastal Expressway, Singapore's first undersea expressway.

Land transportation in numbers (1970-2015).

Year	1970	1988	2006	2015
Cars	150,000	240,000	474,700	602,311
All motor vehicles	260,000	500,000	799,400	957,246
Surfaced roads in km	1,300	2,700	3,262	3,496
Expressways in km	40	90	150	164

Road Networks

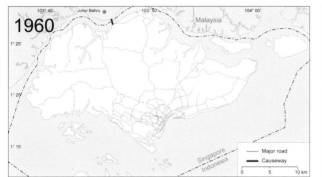

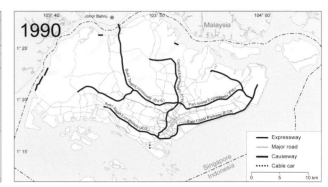

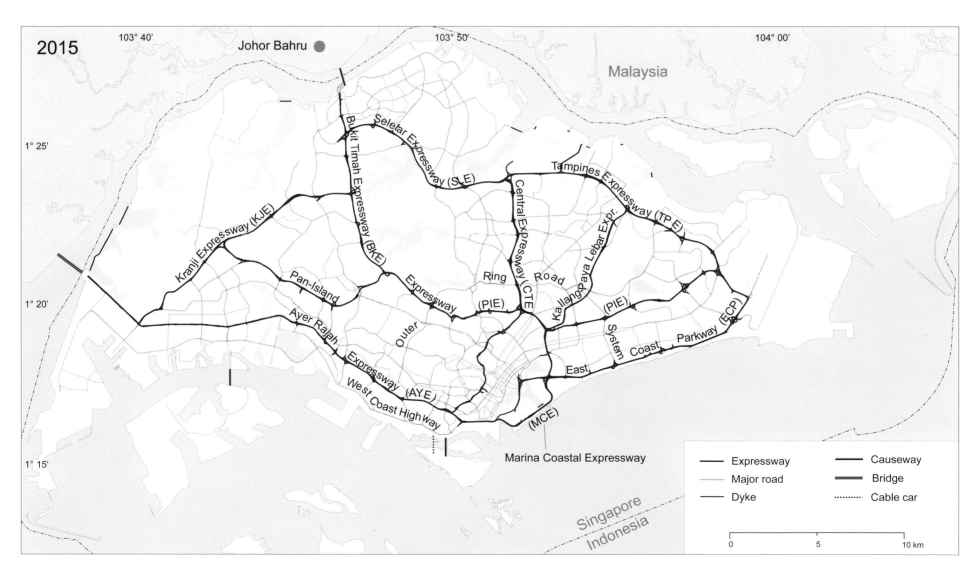

25   Transporting Workers

Fortunately a much denser network of bus lines is well integrated with the train network, thus ensuring that public transport will continue to offer a strong alternative to private car ownership.

The first section of the MRT, between Yio Chu Kang and Toa Payoh on the North-South line, opened on 7 November 1987. The second section, between City Hall and Outram Park on the East-West line, opened on 12 December 1987.

In 1923, a single track metre gauge railway line running from Keppel road railway station in downtown Singapore to the causeway crossing the Johor Strait was built, replacing a line laid down 20 years earlier and further to the east. This very slow "Orient Express" service ceased in July 2011, or more precisely ceased servicing Singapore itself (the link will be revived when the Kuala Lumpur–Singapore High Speed Rail becomes operational in a few years – p. 62). Along with the antiquated yet charming art deco Tanjong Pagar Railway Station, it actually belonged to Malaysia. Apart from an extension to the Jurong Industrial Estate, constructed in the early 1960s and since removed, it had hardly been improved and remained little used, except by a few daily freight trains and two or three short passenger trains running daily and slowly between Singapore and Kuala Lumpur.

However, beginning in the late 1980s, an urban rail network was put into place within the island of Singapore, and has since expanded at record pace. The Mass Rapid Transit (MRT) system's first five stations opened in November 1987. Before the end of 1990, 41 stations and 67 km of rail line were operational. By 2015, the number of MRT stations had reached 90 and the network comprised six lines covering nearly 130 km. In the north and northeast the MRT network linked to an additional 31 km of track handling the so-called Light Rapid Transit (LRT) trains. The LRT network was launched in 1999 and now comprises three lines and 43 stations.

The laying out of the MRT lines went through a series of stages well illustrated by the maps. By 1990, two lines were operational. The first, called the East West Line, traversed the southern part of the island, linked the urban core with the fast industrialising southwest, itself increasingly centered on Jurong, and extended eastwards, eventually to Changi Airport as illustrated by the 2005 map. The second, called the North South Line – which perhaps could be more aptly named the Outer

The MRT Network

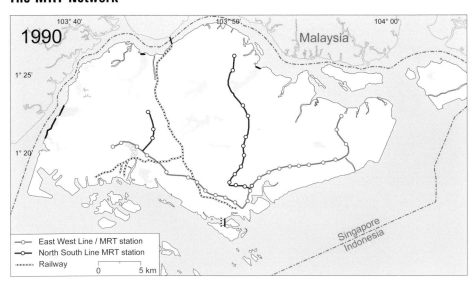

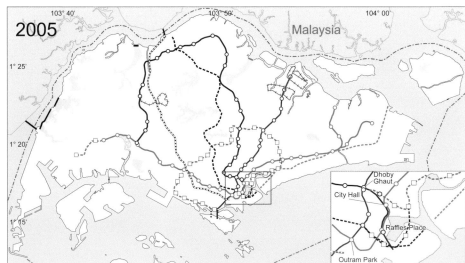

Circle line – links, through a big loop, the central northern parts of the island with its urban core. The 2005 map illustrates this as well as the North South line's links with (1) the East West Line, (2) the North East line which itself links the urban core with the increasingly populous northeast of the island, where two LRT lines redistribute the commuters, and finally (3) the (Inner) Circle Line, which was completed in 2009.

The MRT network has since continued its expansion, with additional lines planned as the demographic filling in of the island's peripheral areas will continue. But notwithstanding the constant development of massive MRT stations, largely relying on the labour of foreign workers, the train network still cannot cover adequately the entire territory, its density remaining much lower than that found in cities such as Paris or London. Fortunately a much denser network of bus lines is well integrated with the train network, thus ensuring that public transport continues to offer a relatively viable alternative to private car ownership.

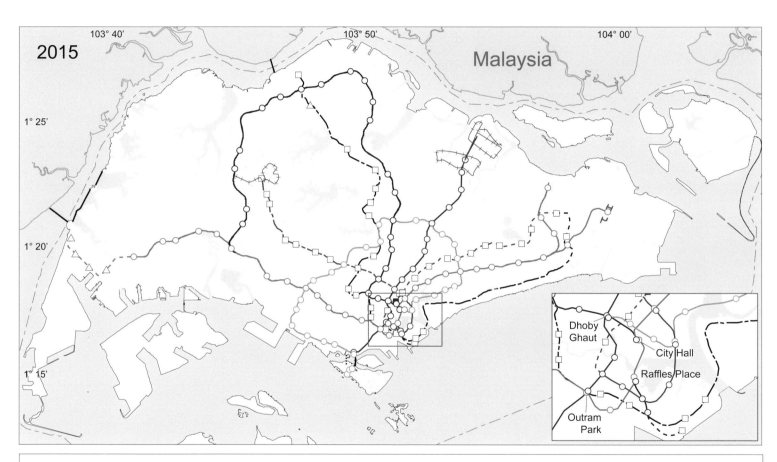

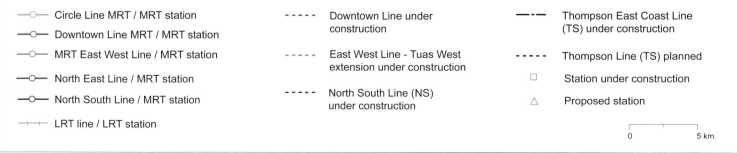

By 2015, the MRT system had five main lines and three LRT lines. By 2030, Singapore expects to add three more lines to the grid.

Two companies currently operate SIngapore's MRT and LRT services:

North East Line
Downtown Line
Sengkang LRT
Punggol LRT

SMRT

North South Line
East West Line
Circle Line
Bukit Panjang LRT

Chapter 5

Services, Control and Entertainment

"The rationale for the Singapore government's approach to nation-building has always been and continues to be the nurturing of the growth of a Singaporean national identity among the population, which will surmount all the chauvinistic and particularistic pulls of the Chinese, Malay, or Indian identities of the various groups on the island."

– Jon S. T. Quah, 1990, p. 45

"The possibly dark side of this policy is the management and control of the internal environment which the PAP government claims are necessary for the successful management of the external environment."

– Linda Y. C. Lim, 1990, p. 135

The relocation of huge numbers of people, largely achieved through the development of New Towns, has been accompanied by the laying out of a new environment, to which these relocated persons must adapt. The reshufflings affect everything from schools and temples, to burial arrangements, to social and political monitoring institutions such as community centres and, finally, to recreational facilities. The continuous changes illustrate the extent to which the entire Singapore living environment is constantly redefined. They also suggest the degree to which the population is at the disposal of the perpetual development process and the tight control that it necessitates.

One of the more striking elements among the overall transformations over the last half-century is the redistribution and sheer proliferation of places to pray – mosques, temples and especially churches. Conversely, cemeteries have seen a drastic reduction in their numbers, with nearly all the remaining ones having been relocated. Speaking of proliferation, that of all levels of educational institutions, particularly higher ones, has taken a complex route. For a time, there was a phenomenal increase in the number of vocational or specialised schools, a trend that has come under control and been regulated. On the other hand, institutions of higher learning and research have been growing steadily and recruiting worldwide.

From the late 1950s onwards, socio-political control has involved the reliance on community centres, although this function has somewhat declined as leisure activities in Singapore have proliferated and become, apparently, an essential mode of social participation. Those activities include the practice of a national "sport" – shopping – for which increasingly extensive malls are endlessly deployed nearly all over the island. Along with more fundamental functions, such as trading and finance, Singapore's partial transformation into a giant supermarket has also contributed to the development of the tourism industry, including medical tourism.

At first surprisingly slow at recognising the need for a true heritage policy, the authorities have been catching up, with the National Heritage Board increasingly active in advocating the preservation of the country's cultural heritage, through the gazetting of national monuments and historic sites. But overall, control of social and political participation continues to rely on systematic regulation, including outdoor eating practices, which is generally well accepted, and constant redefinition of electoral boundaries, which is less.

Go Kart racing at Kallang Park held in conjunction with Singapore's "Tourist Week". In the background are Public Works Department (PWD) Workshops and old Kallang Airport terminal building which serves as headquarters of the People's Association. Ministry of Information and the Arts Collection, courtesy of National Archives of Singapore.

26 Places to Pray

What better can such a state wish if not that its subjects remain primarily concerned with religious and spiritual issues, particularly those with a very materialistic overtone?

Since the late 1950s, the number of places of worship – whether temples, mosques or churches – has increased substantially, although not faster than the overall population. But while the population is now more widely distributed throughout the island, such "places to pray" appear less dispersed. This is due to two factors: the removal of large numbers of families from several peripheral regions, some of which have been completely emptied, and the formation of New Towns, which constitute large population nuclei.

More fundamentally, the evolution in the number of places of worship has been quite different for different groups. For example, by 1988, following expropriations, nearly 90 of the more than 200 Taoist temples existing in 1958 had disappeared, moved or merged, and in some cases reincarnated as Buddhist temples. During that same period, Malay and Indian communities, no less subject to expropriations and resettlement, nonetheless witnessed a substantial increase in the number of their mosques and temples. But this growth in the number of Buddhist, Muslim, Hindu or even Sikh places of worship appeared much less striking than that of Christian churches, which increased in number from 32 to 127. This trend was maintained between 1988 and 2005, with the number of churches nearly doubling from 127 to 239. The same period brought a reinvigoration of Taoism along with Chinese cultural pride, but the trend did not extend to mosques and Hindu and Buddhist temples, whose numbers all decreased.

Since the late 1950s, the number of places of worship – whether temples, mosques or churches – has increased substantially, although not faster than the overall population. But while the population is now more widely distributed throughout the island, such "places to pray" appear less dispersed. This is due to two factors: the removal of large numbers of families from several peripheral regions, some of which have been completely emptied, and the formation of New Towns, which constitute large population nuclei.

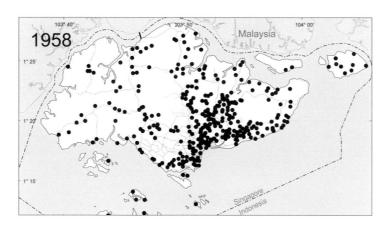

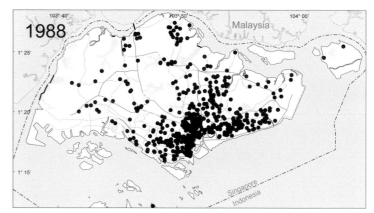

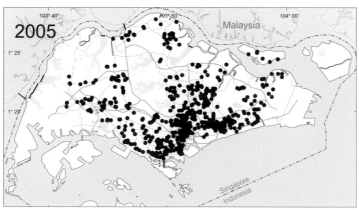

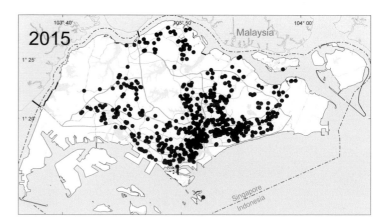

0   5   10 km

● Temple, mosque
   or church

― Major road

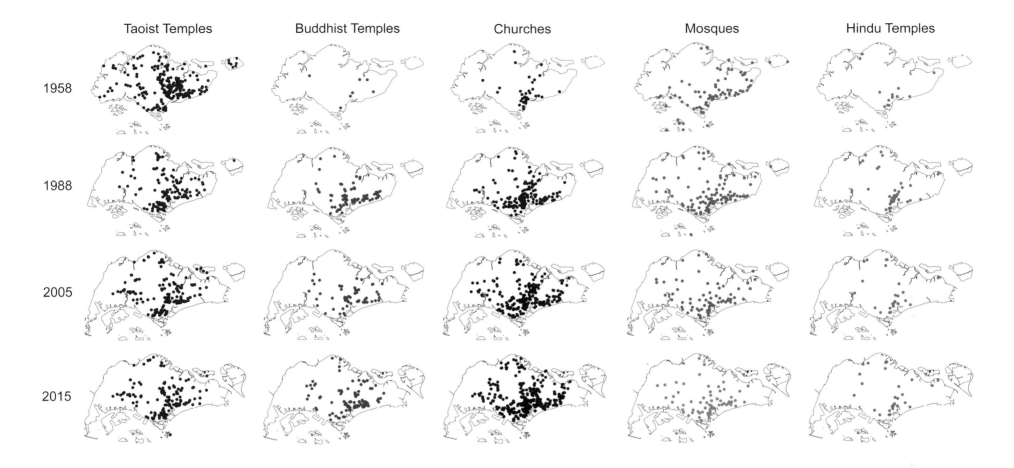

	Taoist Temples	Buddhist Temples	Churches	Mosques	Hindu Temples
1958					
1988					
2005					
2015					

However, between 2005 and 2015, it was the growth rates of Buddhist temples and Christian churches that were by far the strongest. During that decade the number of Buddhist temples nearly doubled, from 62 to 114, while some 120 new churches were established. While churches – whether Catholic or, mostly, Protestant – accounted for less than 10 per cent of all places of worship in 1957, their share had increased to 42 per cent by 2015, with 359 churches out of a total of 745 places of worship. In the meantime, the number of mosques has continued its decline, the most and in fact only significant downward trend since 1958 among all "places to pray". The decline in the number of mosques is offset by the fact that new mosques tend to welcome more worshippers than the older ones. A state-sponsored mosque-building fund spent some S$ 297m on 24 new mosques between 1975 and 2016.

While the increase in the appeal for Christian religious practices, particularly of evangelical denominations, is by now a widespread and well-known phenomenon in Singapore, its political implications are perhaps not yet fully understood. One explanation is that the rise in the appeal of evangelical churches – seen also in other East Asian countries – can be attributed to American-influenced patterns of globalisation as well as to the rise of material prosperity. But this may not tell the whole story. Considering the relative political autonomy of this evolution, and the strongly materialistic and apolitical narrative that characterises most Christian religious denominations, this form of Christianisation seems to work well with the paternalistic Singaporean state. After all, an apolitical approach suits a state that insists on the predominance of material preoccupations and rewards, and which claims, rightfully, to have always been able to provide them.

Places of worship		1958	1988	2005	2015
Taoist Temples		215	127	163	173
Mosques		76	99	71	68
Hindu Temples		13	38	26	32
Churches		32	127	239	359
Buddhist Temples		7	75	62	114
Other	Sikh Temples		9	7	7
	Synagogues		1	2	3
Total		343	476	570	756

27 Places for Burial

By 2014, eight of the ten remaining cemeteries were located, in fact juxtaposed, among the low rolling hills in the Western Water Catchment district; with the only relatively close by settlements being those of foreign workers and military personnel.

The nature and location of burial-grounds as well as sites for ancestor worship are important in most cultures and exceptionally so with the Chinese. Yet Singaporean cemeteries have also been the object of curtailment and massive transfers. Their total number, which in 1958 stood at 113, had been reduced to 64 in 1988, 25 by 2005 and ten in 2014. Exhumations followed by relocations have been most frequent in the central urban area and in the peripheral islands, affecting Chinese and Muslim cemeteries in particular. Although reliance on columbaria and crematoria located in the more densely populated areas did begin to spread rapidly, a sector covering several hectares in the western part of the island has nevertheless been earmarked for cemeteries.

This burial reserve ground in the nearly empty Western Water Catchment district is set among low rolling hills and forms one of the most harmonious landscapes on the island. By 1988, a dozen cemeteries had already been established in the area, at the disposal of those who wished and could afford to give their dead a traditional burial, whether following Chinese, Muslim, Hindu, Buddhist, Jew, Parsi, Bahai or Christian rites. Since then, the overall number of cemeteries has been further reduced, to as few as ten, several ones having been cleared from most of the island's regions, except the northwest and, more importantly the extreme west. By 2014, eight of the ten remaining ones were located, in fact juxtaposed, among the hills in the Western Water Catchment district; with the only relatively close by settlements being those of foreign workers and military personnel. As for columbaria and crematoria, their number which had reached 16 in 2005 had been reduced to eight, with three of them also located amidst the burial grounds of the Western Water Catchment district.

The redistribution of cemeteries in Singapore has over the years involved the digging up of tens of thousands of graves. While in most cases the remains have been cremated and deposited in columbaria, in some instances reburials were carried out in the cemeteries concentrated in the Western Catchment District. There are also instances of cemeteries that have simply been closed, i.e. no new burials being allowed, but with the existing graves remaining untouched. That is the case of the three Chinese cemeteries represented just south of MacRitchie Reservoir on the 2005 map. Known as the Hokkien Association, Ong Clan and Bukit Brown cemeteries, they are officially closed; hence their disappearance from official government documents and from the 2015 map represented here. But they are still visited by descendants of the deceased, as well as by tourists! They have even become, Bukit Brown in particular, emblematic of citizen resistance to top down planning. Plans to close these cemeteries, and to remove the burials in order to make space for the development of highways, have been repeatedly postponed following protest by representatives of clan associations and concerned Singaporeans.

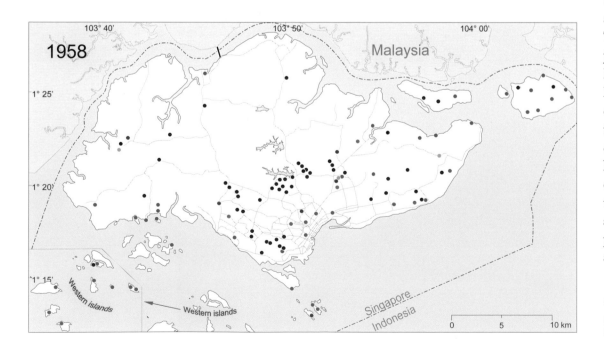

Legend:
- ✦ Columbaria and crematoria
- ● Muslim cemetery
- ● Chinese cemetery
- ● Hindu cemetery
- — Major road
- ● Christian cemetery
- ● Other cemetery
- ○ Closed cemetery

While there were 113 burial grounds in 1958, only 10 remain as demand for space in land-scarce Singapore has meant most have had to be cleared for development.

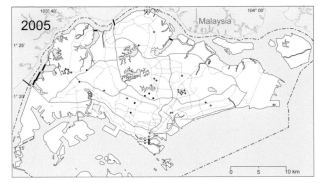

Cemeteries	1958	1988	2005	2014
Chinese	54	25	8	1
Muslim	47	20	8	2
Hindu	3	3	2	1
Christian	3	7	3	1
Others	6	9	4	5
Total	113	64	25	10
Columbaria and Crematoria	0	5	16	7

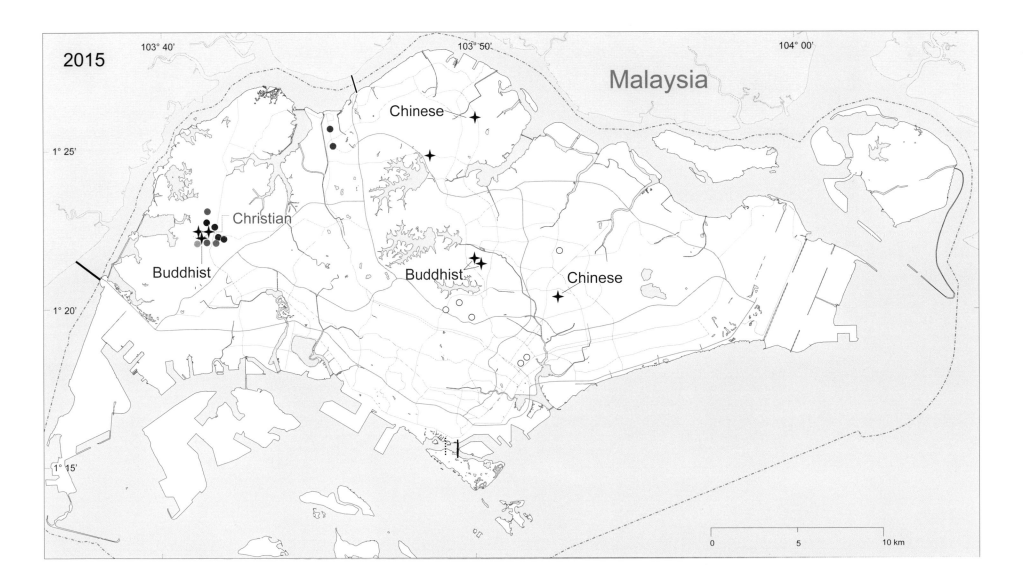

28 Places to Study

The rapid expansion, within Singapore itself, of this network of institutions of higher learning, research centres and labs, particularly striking between 2005 and 2015, is closely linked to the City-State's determination to become a global player in the knowledge field.

While very open-minded on religious matters, the Singapore government's policy on the preferred language for education appears somewhat stricter: the ideal lies in uniformity. Mandarin Chinese, Malay, Tamil, and English are all recognised as official languages but with Malay having the added status of national language, and English being the primary medium of instruction throughout the educational system.

In 1965, schools were identified in ethno-linguistic or religious terms, but such designations no longer exist. Since independence, the Singapore government's commitment to non-communitarian education has never abated and has probably become even more intense over recent years. Even with the very substantial growth of the resources at its disposal, it has continuously allocated at least 10 per cent of the national budget to the education sector, and nearly 25 per cent in the 1970s. School construction has also been quite active.

All levels and denominations included, there were 235 schools in 1965 and 725 in 1990, and the literacy rate climbed from 52 per cent in 1957 to 90 per cent in 1990, 95 per cent in 2005 and nearly 97 per cent in 2015. As expected, the geographical distribution of primary and secondary schools closely follows the distribution of New Towns.

The case of vocational training schools appears particularly interesting. They proliferated during the 1970s and 1980s, mostly in the urban core. Since the early 1990s, their number has been greatly reduced, while the institutions of higher learning, including polytechnics, have steadily increased. Several of these, along with a fast-increasing number of research laboratories, including the National University of Singapore, have been regrouped on its western flank, halfway between the city and the Jurong Industrial Estate, while more recently others have been established at the very core of the City. The rapid expansion of this network of institutions of higher learning, research centres and laboratories in Singapore was particularly striking between 2005 and 2015, and is closely linked to the city-state's determination to become a global player in the knowledge business. It also has an increasing impact on the built environment of the island, several of these institutions becoming architectural landmarks.

Schools	1966	1990	2005	2015
Chinese Schools	107	0	0	0
Malay Schools	37	0	0	0
Tamil Schools	41	0	0	0
English schools	12	0	0	0
Primary and Secondary Schools		316	357	339
Institutions of Higher Learning	6	8	16	57
Other schools	79	405	130	265
Total	282	729	503	661

Schools and Institutions of Higher Learning

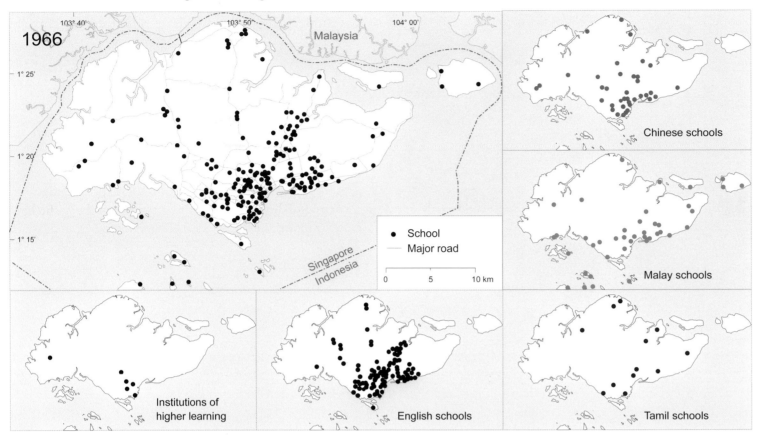

Education in pre-independent Singapore was conducted in different languages, hence different schools catered to their own community's needs. In 1960, a policy of bilingualism in schools was officially introduced.

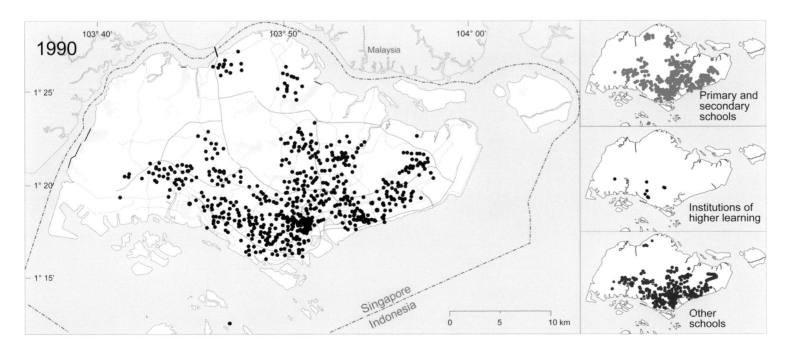

The phasing out of Chinese, Malay and Tamil schools began in the 1970s. By the 1990s, all primary and secondary schools came under the purview of the Ministry of Education.

Even if English is the premium language of instruction as well as the lingua franca of Singapore, members of the Chinese community have become increasingly fluent in Mandarin, in striking contrast with what prevailed at the time of independence.

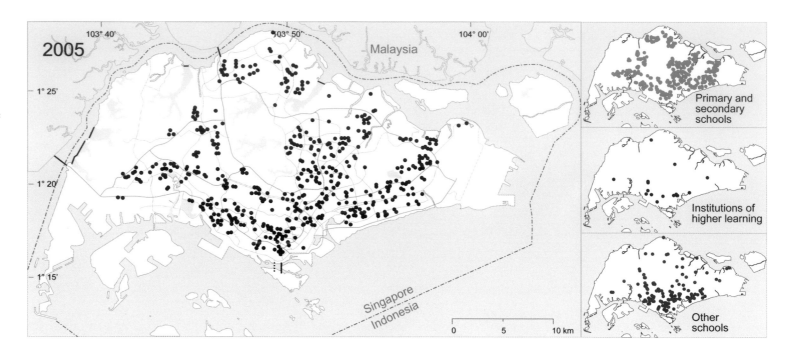

In the past 15 years, the number of universities has risen sharply and so have the resources put at their disposal. The Singapore Management University opened in 2000, the Singapore Institute of Technology started operations in 2010, and the Singapore University of Technology and Design opened in April 2012,

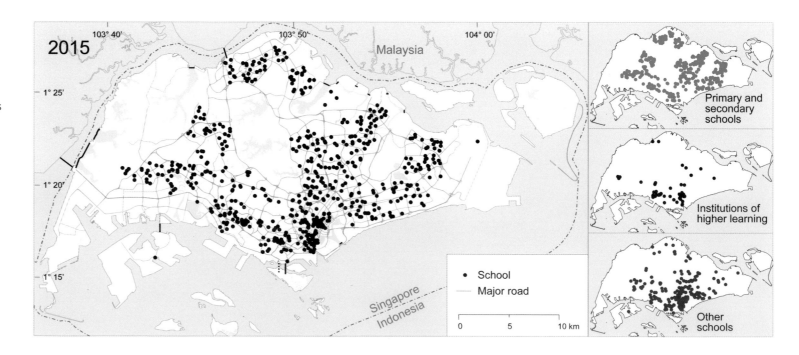

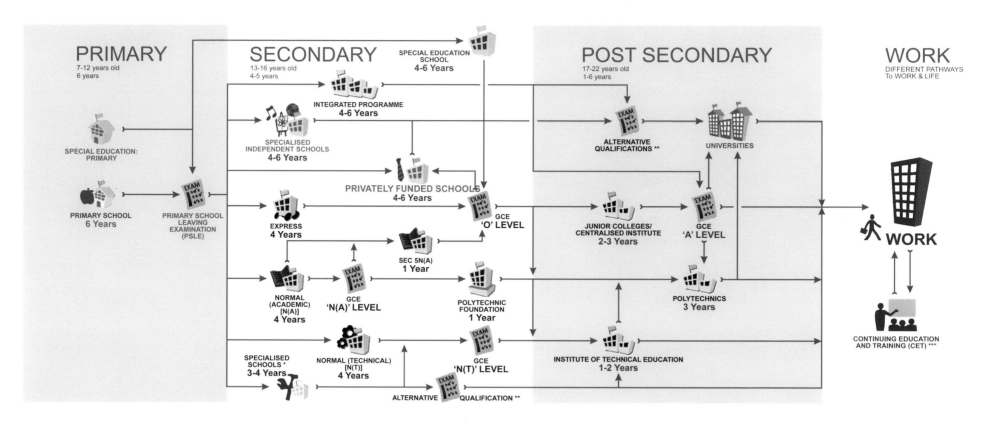

PRIMARY
7-12 years old
6 years

SECONDARY
13-16 years old
4-5 years

SPECIAL EDUCATION
SCHOOL
4-6 Years

POST SECONDARY
17-22 years old
1-6 years

WORK
DIFFERENT PATHWAYS
To WORK & LIFE

SPECIAL EDUCATION:
PRIMARY

PRIMARY SCHOOL
6 Years

PRIMARY SCHOOL
LEAVING
EXAMINATION
(PSLE)

INTEGRATED PROGRAMME
4-6 Years

SPECIALISED
INDEPENDENT SCHOOLS
4-6 Years

PRIVATELY FUNDED SCHOOLS
4-6 Years

EXPRESS
4 Years

GCE
'O' LEVEL

ALTERNATIVE
QUALIFICATIONS **

UNIVERSITIES

JUNIOR COLLEGES/
CENTRALISED INSTITUTE
2-3 Years

GCE
'A' LEVEL

WORK

NORMAL
(ACADEMIC)
[N(A)]
4 Years

GCE
'N(A)' LEVEL

SEC 5N(A)
1 Year

POLYTECHNIC
FOUNDATION
1 Year

POLYTECHNICS
3 Years

SPECIALISED
SCHOOLS *
3-4 Years

NORMAL (TECHNICAL)
[N(T)]
4 Years

GCE
'N(T)' LEVEL

INSTITUTE OF TECHNICAL EDUCATION
1-2 Years

CONTINUING EDUCATION
AND TRAINING (CET) ***

ALTERNATIVE   QUALIFICATION **

Singapore's education
system. Courtesy of the
Ministry of Education*
Specialised schools offer
customised programmes for
students who are inclined
towards hands-on and
practical learning. These
schools include Northlight
School, Assumption Pathways
Schools, Crest Secondary
School and Spectra
Secondary School, which
opened in 2014.

ITE College West in Choa
Chu Kang provides a post-
secondary education in the
form of pre-employment
training to secondary school
leavers and continuing
education and training to
working adults.

** Alternative Qualifications
refer to qualifications not
offered by the majority
of mainstream schools in
Singapore.
*** Continuing Education and
Training (CET) is designed for
adult learners or companies
looking to upgrade the skills
and knowledge of their
employees.

29   Places for Recreation

The success of Pulau Sentosa, which can be reached by ferry, cable car or over a causeway, has been such that Pulau Ubin, in the Strait of Johor, is being transformed into another type of recreation platform.

Towards the end of the 1950s, forested areas and natural beaches, particularly in the outer islands, were among the favourite recreational areas of Singaporeans. As access to this "wilderness" was progressively curtailed by the depletion of the natural forest and the industrialisation of the shoreline, enormous energy went into the development of parks, beaches, stadiums and swimming pool complexes – in principle, each New Town has an Olympic-size pool – and other sports facilities. Parks, which do represent major places for recreation, have been discussed above in connection with the evolution of the Garden City (p. 32).

To these green parks could be added numerous leisure attractions – not represented here – that have sprung up in several areas. These include the Jurong Bird Park, and the extensive Zoological Garden near the Upper Seletar Reservoir. Equally noticeable, the establishment of specialised recreation sites has brought about the thorough transformation of substantial pieces of land. This is the case, notably, of Pulau Sentosa, the "island of tranquility". Located a short distance from the urban core, just beyond Pulau Brani, it is often referred to, and for good reason, as the Disneyland of Singapore. Clearly, it is not exactly tranquil. This island was one of the choice pieces in the land bank ceded to the Singapore government by the British military following their 1968 to 1971 withdrawal. Since then, it has been extended and equipped to host a wide range of recreational facilities, including theme parks, museums, resort-type hotels, marinas,

a very prominent golf course, and beaches largely fabricated with imported sand and cement boulders made to look like natural ones. The success of Pulau Sentosa, which can be reached by ferry, cable car or a causeway, has been such that Pulau Ubin, in the Strait of Johor, is being transformed into another type of recreation platform.

As for golf courses, which attract rich tourists but also rich Singaporeans, in 2005 their number had reached 23 and they occupied nearly 17 square km or more than 2 per cent of the national territory, an astonishing proportion for a land hungry city-state. This "luxury" did not remain unnoticed among Singaporean planners. The number of golf courses has since been cut to 17 and their total extent reduced to some 12 square km. But the number of large stadiums and swimming pool complexes has continued to increase, a commendable achievement, as these are public goods and accessible to all.

Universal Studios Singapore is one of the newest major attractions in Singapore. Located within Resorts World Singapore on Sentosa, it is the first Universal Studios theme park in Southeast Asia.

Over the years, the number of public recreation facilities available for residents has increased steadily, while the number of private golf courses has been curtailed.

Places for Recreation	1958	1988	2005	2015
Golf Courses	2	13	23	17
Public Swimming Pools/Complexes	3	23	22	25
Stadiums	0	13	17	21

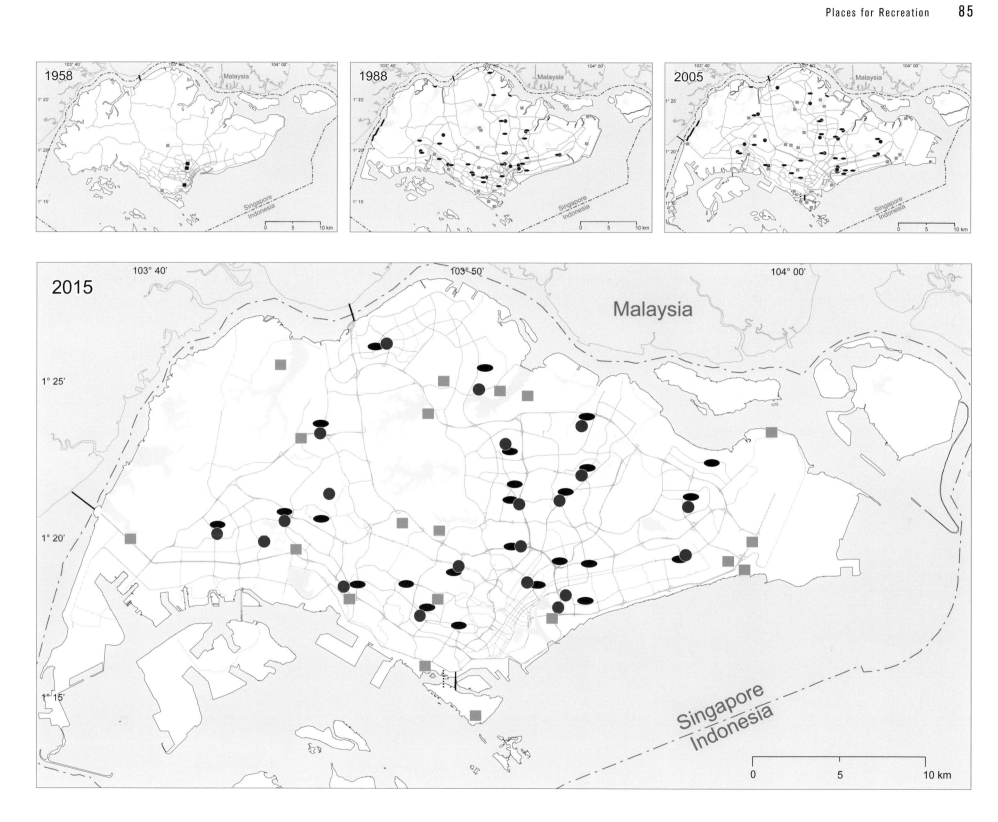

1958

1988

2005

2015

Golf course   Public swimming pool   Swimming complex   Stadium   — Major road

30 Rallying Points: Community Centres

Community Centres seem to have maintained their pertinence, their redeployment following that of the population, as apparent, for example, in the northeast with the expansion of the New Towns of Sengkang and Punggol.

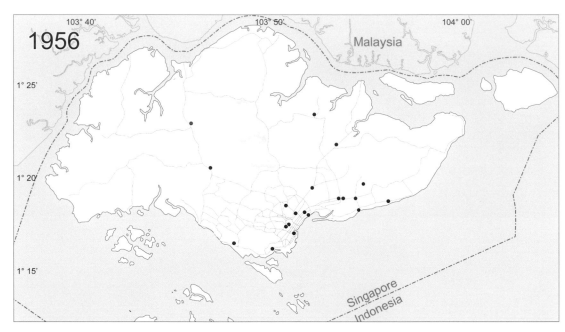

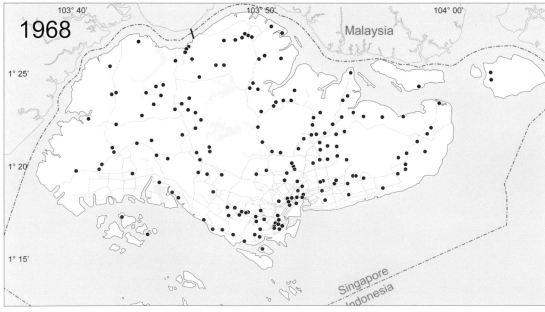

In the mid-1950s there were around 20 Community Centres in Singapore, most of them in the central urban area. Shortly after coming to power in 1959, the People's Action Party (PAP) set about increasing their numbers and expanding their functions, until then predominantly recreational. The task was assigned to the People's Association (PA), formed in 1960. Since then, the association has played a crucial role in the population's cultural, social and political nurturing. The latter mandate appeared dominant during the sixties, the centres essentially fulfilling the role of local branch offices or rallying points for the ruling party. Ordinary citizens visiting them for essential services were also exposed to a political message.

Combined with their broad geographical spread, the Centres' proliferation ensured a near total coverage of the territory, whether urban or rural, including, then, the small islands. Today, their socio-cultural function appears prevalent once again; through them people gain access to a broad range of educational and recreational resources, to which are added nursery and kindergarten services.

The location of Community Centres has evolved along with the redistribution of population. Many have been closed down or regrouped, and new ones established. As a result, the distribution pattern of the 112 centres that still existed in 2005 was quite different from that of 1968, when the number stood at 181. The substantial reduction in their numbers that occurred during that period may have implied that their social and political monitoring functions had become less important. However, since then the rate of attrition has slowed down considerably as they were still 106 of them in 2015. They seem to have maintained their relevance, their redeployment following that of the population, as apparent, for example, in the northeast with the expansion of the New Towns of Sengkang and Punggol. The remaining ones have also had the possibility to concentrate and adapt their services to the new forms of population concentration that the New Towns constitute.

	1956	1968	2005	2015
Community Centres	20	181	112	106

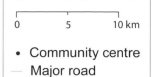

- • Community centre
- — Major road

Designed by William Lim Associates, the postmodern Marine Parade Community Building has a distinct mural cladding called the "Texturefulness of Life", the largest mural in Singapore.

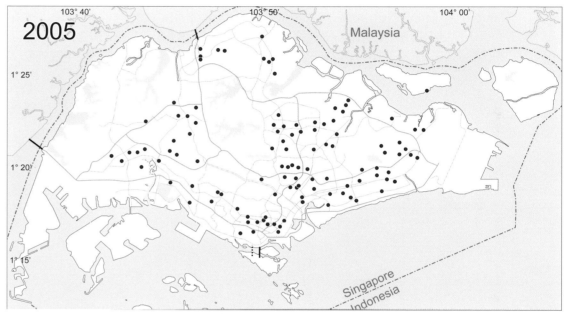

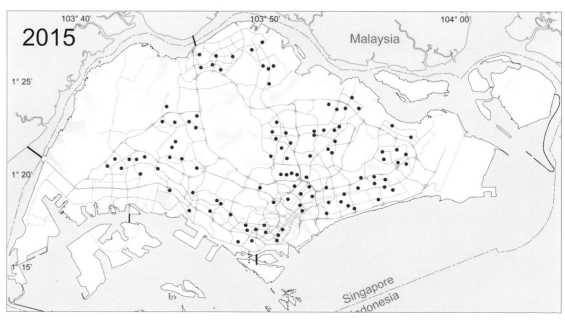

31 Rallying Points: Shopping Malls

Singaporean shopping malls have become, to a degree rarely seen elsewhere, essential cultural and social rallying points.

The development of shopping plazas or malls has closely accompanied Singapore's economic growth as well as its territorial overhaul. As the economy grew and the population's affluence increased, so did consumerism. Over the years, spending leisure time in increasingly large shopping complexes has become a modus vivendi for a majority of Singaporeans. Not only have shopping malls been deployed within or near every New Town, the opening of each MRT station usually involves the development of rows of new shops, within or around its precincts.

Throughout Singapore, total retail space was multiplied by more than six between 1967 and 2014 (while overall population, of permanent as well as non-permanent residents, was multiplied by less than three), with a striking acceleration from the early 1980s onwards. As for shopping plazas per se, by 2015, 164 can be identified, with the total number of retail outlets found on their premises probably reaching 20,000. The largest concentrations of plazas are still found in the

downtown area, which has maintained its function as the leading retail trading domain, still hosting nearly a third of all retail outlets in the country, including in the fast developing HarbourFront area, with, for example the VivoCity shopping mall, the largest in the country. And, notwithstanding the redistribution of population and the development of New Towns, Orchard Road has not lost its own dominance within the urban centre as well as within the island. Its numerous plazas, with many being periodically refurbished or upgraded, still represent the central magnet, for both foreign tourists and perhaps even more local residents, particularly citizens (p. 90).

This said, as the outlying urban nuclei are increasing their overall autonomy some huge shopping malls have also expanded within their precincts. Consider, for example, the case of Jurong Point 1 and 2, with 70,000 square meters at the disposal of 450 shops, visited by more than three million shoppers per month! Online shopping may be taking business away from retail shops and even more from department stores, but shopping plazas are hard to beat for socialising. As with markets in traditional societies, Singaporean shopping malls have become, to a degree rarely seen elsewhere, essential cultural and social rallying points.

Retail Space, 1967–2014

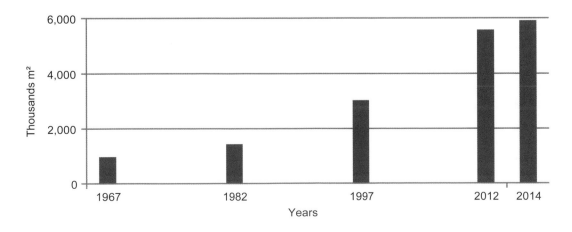

VivoCity, the largest shopping mall in Singapore, was designed by Japanese architect Toyo Ito and is located at Harbourfront. Photograph by Sengkang, distributed under a CC-BY 2.0 license.

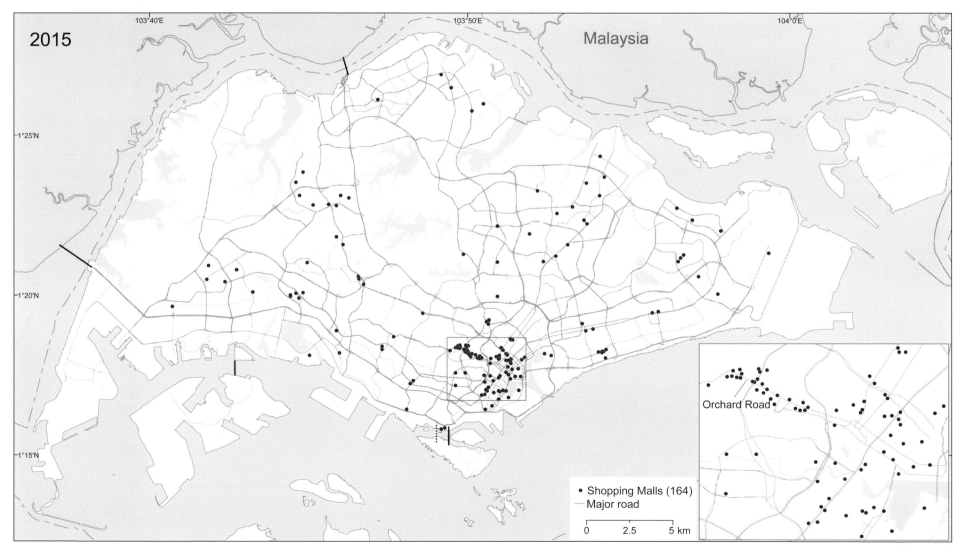

32 Orchard Road: The Shopping cum Tourist Trail

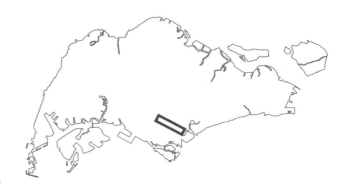

"It is a world-famous shopping precinct, a tree lined retail haven that has attracted locals and tourists alike for decades. But there are concerns that a slowing economy is taking the shine off Orchard Road. However, the shopping super-strip can still stay streets ahead."

– Straits Times, 15 July 2015, p. B 9

"[On Orchard Road], for the rest of the time, pedestrians are forced underground to make more room for cars."

– Bruno Wildermuth, 2015, p. 18

Along with skyscrapers bordering the Singapore River, hotel towers are now a characteristic feature of Singapore's urban profile (p. 92). To make room for them and the numerous shopping plazas that delight tourists, foreign residents as well as Singaporeans, a major avenue and its surroundings were redesigned.

Since the late 1960s and early 1970s, Orchard Road and its extension towards Marina Bay have been made into a kind of shopping cum tourist trail. By itself, that 5 km trail hosts more than 50 hotels. Twenty-five of them are located along Orchard Road or in its immediate vicinity and offer a total of more than 10,000 rooms. To accommodate these hotels and the equally ubiquitous and even more extended shopping plazas, the road network was redesigned and augmented, another routine operation in contemporary Singapore. This overhaul was largely accomplished during the 1970s and 1980s, with Orchard Road remaining the central axis. But its appearance was thoroughly transformed with shopping plazas and hotels crowding in on either side.

Since then the pace of change has somewhat levelled off, but upgrading and changing functions keep occurring, with some high-rise buildings, such as the Mandarin Orchard Singapore hotel, having become permanent landmarks. So have two huge and more recent multi-storied shopping malls, Ngee Ann City and Ion Orchard, the latter spreading over some 66,000 square meters of floor space and boasting more than 300 retail food and beverage and entertainment stores, with Orchard MRT station buried beneath the mountain of shops!

Increasingly crowded, now infamous for its slow traffic as well as occasional traffic jams, and surprisingly little concern for pedestrians, Orchard Road faces fierce competition from other huge shopping complexes located in or near several of the New Towns or urban nuclei, such as Jurong Town. But, for the moment, Orchard Road still succeeds in maintaining its buzz, by overhauling not so much its layout but the design of its shopping world. Often but somewhat inappropriately referred to as the Singaporean "Champs-Élysées" – the grand Parisian avenue that is much wider, more pedestrian friendly and links major historical and architectural landmarks – Singapore's premier shopping cum tourist trail has nevertheless become a landmark of its own.

The first shop of note established on Orchard Road was Tangs in 1958. Later on, in 1974, Orchard Road became a one-way street after more than a century as a two-way thoroughfare. The new North-East line (2003) and Circle Line (2010) at Dhoby Ghaut also made the area more accessible to residents and tourists.

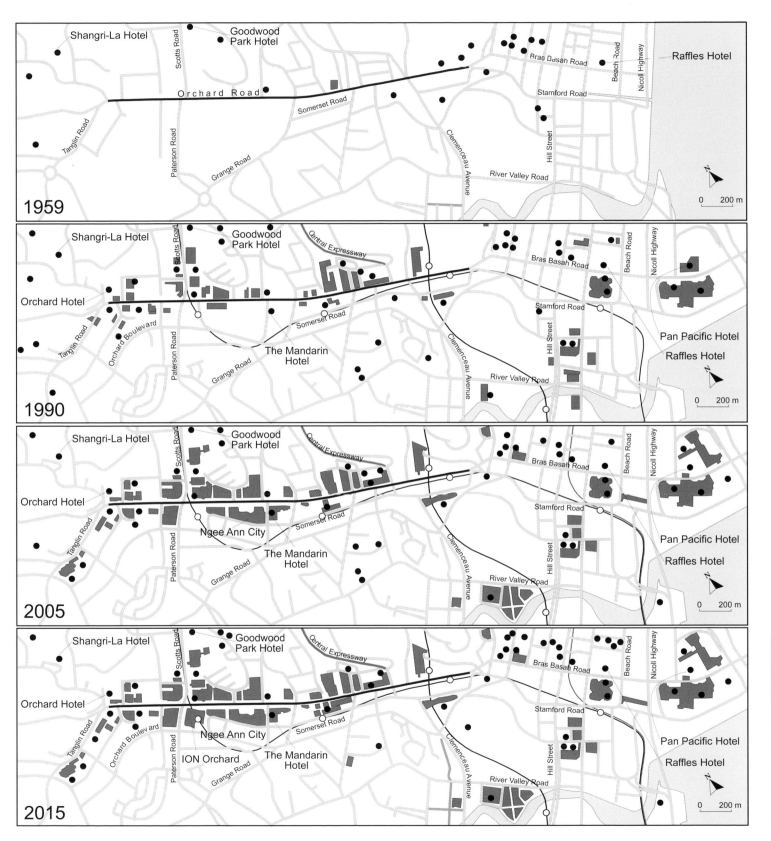

	Shopping Plaza
●	Hotel
▬	Orchard Road
	Road
—○—	North East MRT Line / MRT Station
—○—	North South MRT Line / MRT Station
——	East West MRT Line

33 Hosting Foreign Visitors

This strong rate of growth in the number of hotel rooms is largely attributable to Singapore having positioned itself to become a prime destination for business people and well-off tourists.

In the late 1950s, Singapore did not appear particularly attractive to travellers, and its tourism infrastructure, apart from a few grand and exotic hotels such as the Raffles or the Goodwood Park, was poorly developed. The fifty-odd hotels, more than half of which were operating without a license, offered a total of about 1,000 rooms; this was when fewer than 100,000 foreigners visited Singapore annually. However, the transformations to the urban landscape and the improvement of transport infrastructure were followed by tourist promotion campaigns and the construction of luxury hotels. As a result, by the early 1990s, the city-state had 91 licensed hotels, offering more than 25,000 rooms, with 26 of these hotels each boasting 400 rooms or more. Tourism had become big business and the construction of huge luxury hotels almost a routine, still ongoing, operation. In 2015,

the total number of licensed hotels had reached nearly 400, renting out a total of close to 60,000 rooms!

Initially, hotel expansion centered on Orchard Road and its surrounding streets. It has since spread throughout much of the downtown area, along the Singapore River banks, particularly the south bank, amidst banking towers, into Chinatown, and beyond, noticeably on the richly endowed and massively transformed resort island of Sentosa with its 13 hotels, several of which actively attract Singaporean guests. Hotels have also expanded eastwards, well into the Geylang area. Budget hotels, which are common in this area, are not typical of the Singapore hotel business.

Early on, when it chose to open the door to international tourism, the Singapore authorities and in particular the Singapore Tourism Board were explicitly aiming at wealthier travellers and more specifically at business travellers: hence the predominance of large luxury hotels. If some of these, such as the Raffles and Goodwood Park, are truly heritage landmarks, the local specialty has become huge luxurious hotels, such as the Marina Bay Sands. Opened in 2010, the Marina Bay Sands has probably become Singapore's most renowned architectural landmark, with its three towers linked at the top by a boat-shaped terrace. Given its location near Marina Bay, the 57 stories and 200 metre-high structure is visible from the sea and from numerous locations on the island. It offers 2,560 rooms and, on the terrace, according to

Licensed Hotels

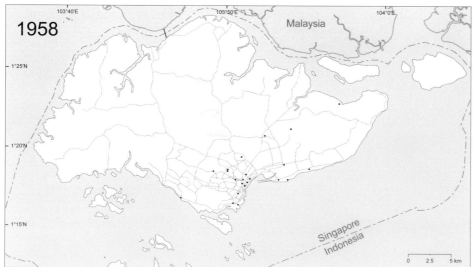

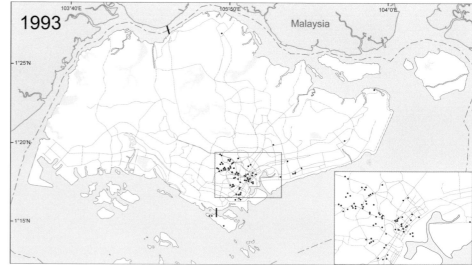

its own publicity "the hotel guests enjoy the exclusive access to the highest and largest infinity pool in the world". Finally, back on the ground, the Marina Bay Sands hosts on its precincts a casino, as well as the ArtScience museum.

Overall, the rate of growth in the number of hotel rooms has been strong and steady and even if the increase in revenue did taper off during the financially difficult late 1990s and early 2000s, both number of rooms and revenue have enjoyed strong growth. This is largely attributable to Singapore having positioned itself to become a prime destination for business people and well-off tourists, the number of international visitor arrivals having in 2014 reached 15 million. To this must be added the government's policies to attract medical tourism (p. 94) and to encourage residents to stay in local hotels during weekends, when business tourism slackens. In 2014, about 75 per cent of foreign visitors were from Asian countries, with, in the lead, Indonesian (20 per cent), Chinese (11 per cent) and Malaysian (8 per cent) ones.

Despite the general upward trend in hotel revenue, a hotel room glut meant that September 2016 saw revenue per available room drop to its lowest levels since 2010.

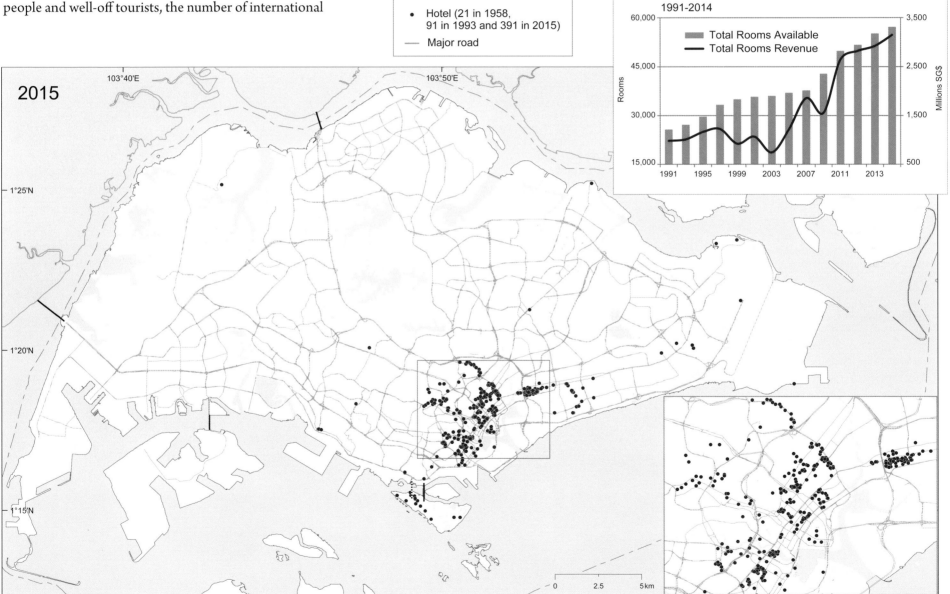

Legend:
- Hotel (21 in 1958, 91 in 1993 and 391 in 2015)
— Major road

1991-2014
- Total Rooms Available
- Total Rooms Revenue

2015

34 Taking Care of People's Health

"Singapore has one of the most successful healthcare systems in the world, in terms of both efficiency in financing and the results achieved in community health outcomes."

– John Tucci, 2004

In 2014, Bloomberg ranked Singapore's healthcare system number one in the world, ahead of Hong Kong, Italy and Japan. The city-state was also ranked ahead of these and other leading countries with regard to healthcare cost as percentage of GDP. That percentage stood at 4.5 per cent, compared to 5.3, 9.0 and 10.2 per cent respectively for the other top three (the USA, ranked 44th, it reached 17 per cent). How did the city-state get there?

Early on in the independent republic's history, the government gave priority to housing, education and, of course, healthcare. This meant, first, progressively devising a complex yet efficient health funding programme relying on individual compulsory savings through payroll deductions, state subsidies within a nationalised health insurance plan and price controls. Up to 80 per cent of Singaporeans obtain their medical care exclusively within the public health system, although the provision of private medical services is increasingly offered to the well-to-do locals and equally

well-off foreigners who travel to Singapore specifically to receive high quality and efficient medical treatment. In fact, Singapore actively encourages so-called "medical tourism" as it represents a substantial source of funding for the island Republic's medical sector.

Secondly, improvement of health services relied on the deployment of a large network of clinics that has closely followed the geographical redistribution of population. These clinics offer a Community Health Care Assist Scheme that enables Singapore citizens from lower- and middle-income groups to receive subsidised medical and dental care.

For lack of reliable data concerning the years prior to the 1990s, it is difficult to document the early stages of this deployment. But it is obvious that frontline medical services are today available throughout the populated regions of the island, with the exception of those where foreign workers' dormitories have recently been established, notably in the western reaches of the island. As for hospitals and hospital beds, their number has doubled between 1958 and 2015, from 17 to 34, hence at a rate of growth slower than that of the island's population (p. 40). Obviously, the 698 clinics are making a difference.

In the 2013 Budget, the Singapore government announced that more healthcare institutions would be built by 2030, including four more public hospitals and up to 12 polyclinics.

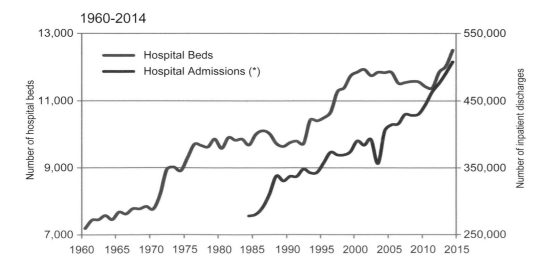

1960-2014

* refers to inpatient discharges for all hospitals. Prior to 2002, data on public sector hospitals refer to admissions.

- Hospital (17 in 1958, 24 in 1993 and 34 in 2015)
- Clinic with Community Health Assist Scheme (698 in 2015)
- Major road

Hospitals and Polyclinics

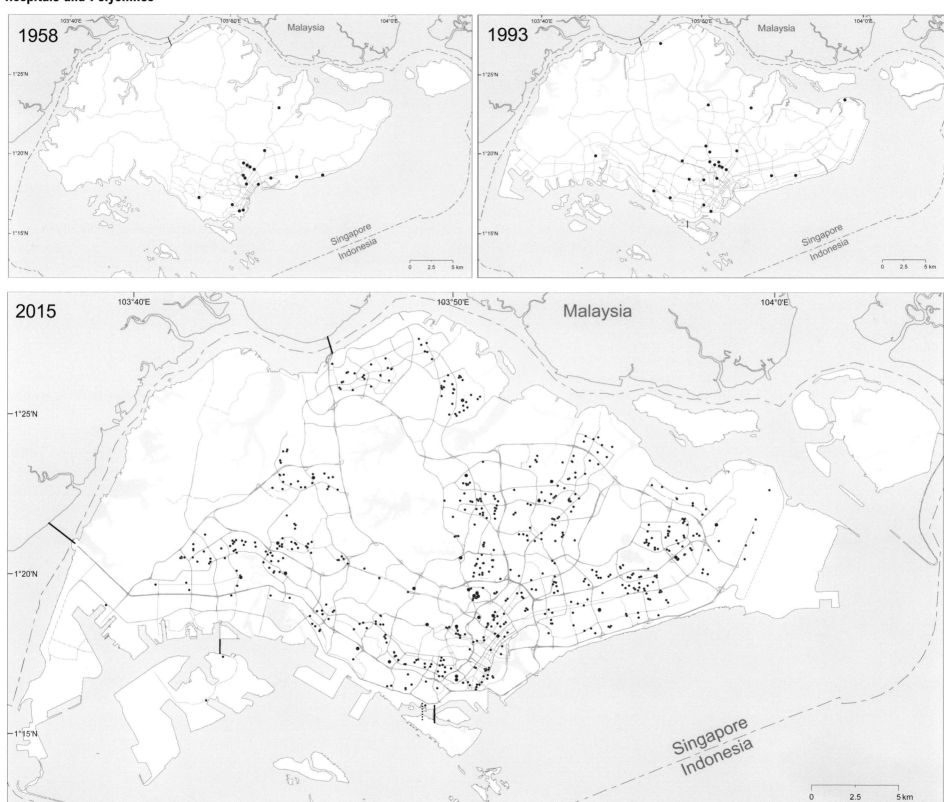

1958

1993

2015

Malaysia

Singapore
Indonesia

35 Honouring the Past

"Another driving force has emerged in Asia, that of National Identity."

– *Tay Kheng Soon & Akitek Tenggara, 1997, p. 45*

Singapore's cultural heritage was seriously threatened during the first two decades that followed independence, as the island was increasingly being turned upside down and renovation of urban sites often implied total demolition and replacement, rather than restoration. The various ministries and boards involved in designating land use, and hence expropriating and relocating people as well as moving earth, paid limited attention to cultural attributes of specific neighbourhoods and places. Nevertheless, as early as 1971, a Preservation of Monuments Board was created under the Ministry of National Development. But it was not very active and there was some amount of civic protest at the apparent cultural indifference with which the territorial overhaul was taking place. By the mid-1980s, at least one non-government, non-profit institution had become quite active and, most importantly, visible.

Founded in 1986, the Singapore Heritage Society has since been successful at fostering awareness about the country's

historical and cultural heritage. It also contributed to raising preoccupations among the various town planning institutions about the national cultural heritage, beginning with Chinatown, the National Library building, Bukit Brown cemetery (Plate 27) and the former Malayan Railway corridor. The Urban Redevelopment Authority (URA) is planning to turn the latter into a green corridor.

But of course, the one institution that bears the responsibility to ensure the preservation of the said cultural heritage is the National Heritage Board (NHB). Since 1993, when its belated confirmation as the one coordinating body responsible for the protection of the national cultural heritage, it has been catching up and has gazetted an increasing number of national monuments. While there were only eight in 1979 and only 23 in 1990, the number of national monuments had reached 70 by 2015. Perhaps more significant, since its creation the NHB has launched into a vast programme of recognition of "marked historic sites" whose number stands at 276 in 2015. While the former are necessarily structures, such as religious monuments which account for 26 of the 70 officially recognised ones, "sites" can be the same but also simply locations of historical events or associated with a personality having played a significant role in the nation's history. These sites are sometimes marked by a simple plaque, many being dedicated to WWII events. Their distribution throughout the island is also less concentrated in the urban core than that of monuments.

National Monuments and Heritage Sites

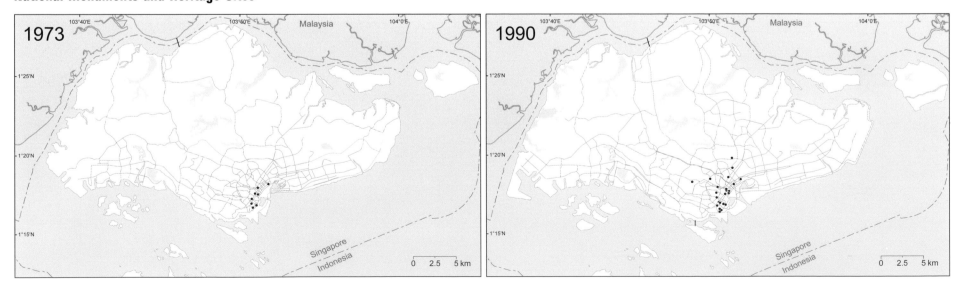

The same can be said of the 14 Heritage Trails that have been developed by the NHB with the participation of local communities including schools. Overall, these represent a still modest but relevant attempt at capturing some key elements of Singapore's national identity

National Monuments, 1973–2015

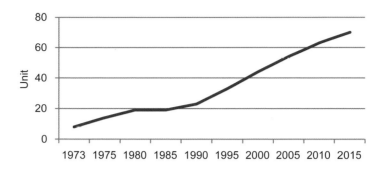

The early monuments were based around the downtown core area, especially around the civic district and Raffles Place area. As the notion of heritage expands in Singapore, the types of monuments and sites have likewise broadened.

- National monuments (8 in 1973, 23 in 1990 and 70 in 2015)
- Heritage sites (276 in 2015)
- Major road

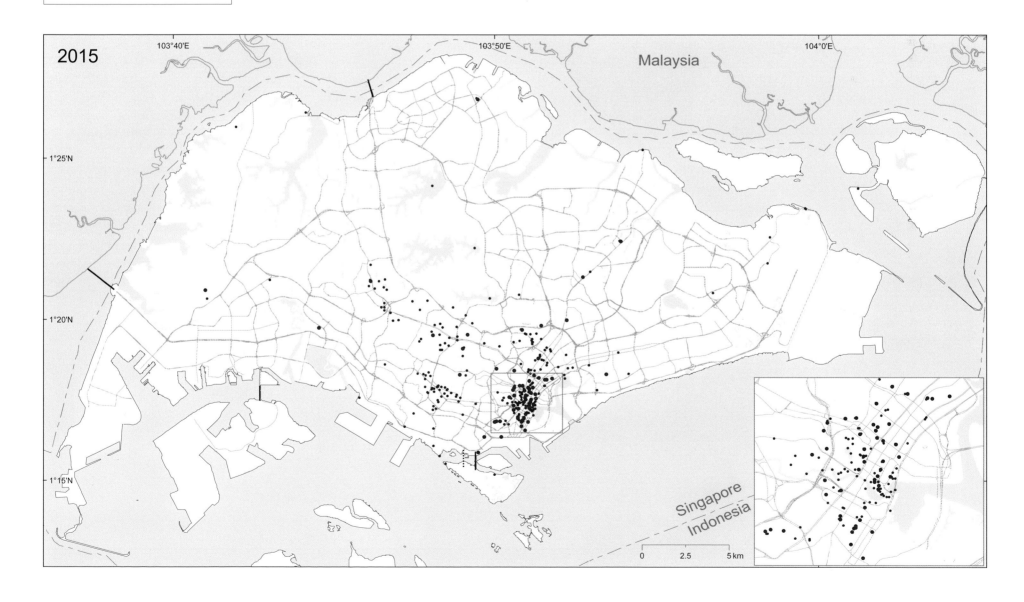

36 Regulating Outdoor Gastronomy

"These everyday places have become Singapore icons etched into the cityscape. [Hawker centres] are a microcosm of [...] society, and have mirrored the changing life and landscape of Singapore over time."

– Lily Kong, 2007, p. 19

The post-war and post-independence history of food vendors in Singapore is in many ways illustrative of the nature of the City-State's permanent transformation, involving uprooting, relocation and upgrading. Closely tied to Singapore's history, eating out, in sheltered as well as unsheltered premises, has long been a national past-time, with the food served in these precincts as well as the caterers (i.e., the food vendors) occupying a special place in the national history.

In the 1950s and 1960s, itinerant and not so itinerant street vendors were a familiar sight. For example, among families living in the small and cramped apartments of Chinatown, cooking and eating in the street was common. Lack of space in their tiny one or two-room flats as well as the possibility or need to sell specialty dishes to neighbours warranted it.

The Maxwell Road Food Centre was established in the 1980s, originally as a temporary site for hawkers displaced from elsewhere in Chinatown, and on the site of a 1950s wet market. It was rebuilt in 2001 and remains one of Singapore's favourite. Photo by David Berkowitz (Flickr user davidberkowitz) shared on a CC-BY 2.0 license.

This led to or was associated with the practice of roaming or itinerant food vending. Given the high rate of unemployment that prevailed in the 1950 and 1960s, this informal employment had become attractive to many looking for an income. As a result, the number of itinerant food hawkers increased, having allegedly reached 24,000 by the mid-1960s. However, this unregulated expansion brought about serious hygiene problems, even food poisoning, as well as congestion in streets and public spaces. Notwithstanding their popularity with the public to which they provided cheap and generally tasty food, hawkers had to be brought under legal regulation (i.e., given licences) and settled down in specific locations.

Initially under the responsibility of the Hawkers Department, this process was accelerated during the late 1960s and, from 1971 to 1986, a well-funded programme was implemented to establish hawker centres. These were often associated with outdoor markets, with which they could share necessary infrastructure and services, such as shelter, clean water, drainage, etc. The aim was to resettle some 18,000 hawkers and the task given to the Housing and Development Board (HDB). The Urban Redevelopment Authority (URA) and the Hawkers Department were also involved, the latter being allocated, by the Land Office, with sites on which to build hawker centres, whose fundamental and definitional characteristic remain that they are open-air structures.

So, here again, the Singapore government's status as sole owner of land was to facilitate eviction and relocation. In 1986, by which time no illegal hawkers remained in Singapore, more than 6,000 of them were operating, under licenses, in some 140 hawker centres. But since then, a number of these centres have either disappeared or been regrouped. According to the National Environmental Agency

(NEA), which now manages and regulates the remaining 107 centres through a typically Singaporean Hawker Centres Upgrading programme (HUP), the numbers for both licensed hawkers and hawker centres dropped significantly until 2003. But, in 2004, those for licensed hawkers rebounded spectacularly, presumably due to a modification in the legislation or in the way statistics were tabulated. In fact, 2004 was the year when the NEA was given exclusive responsibility over the management of hawker centres, taking definitely over all other agencies still involved such as the HDB.

Meanwhile, closed-in and massively air-conditioned "food courts" have multiplied, in close association with the proliferation of crowded and equally frosty shopping malls (p. 88). These generally more expensive food courts do compete with hawker centres but are often considered less appealing as far as local and particularly gastronomic colour and culture is concerned! After all, hawker centres are among the few remaining living heritage sites or institutions on an island involved in a permanent whirlwind of cultural and environmental change and regulation.

Hawker Centres

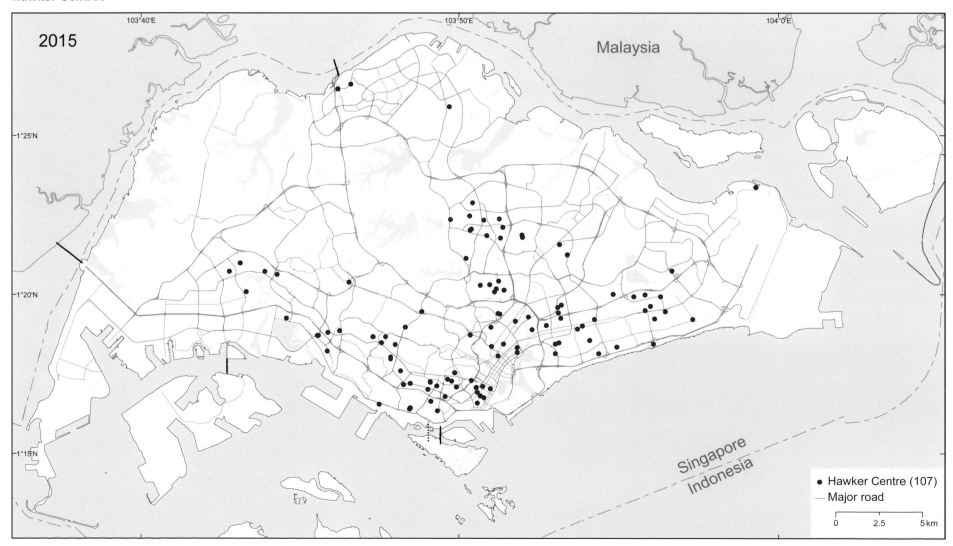

37 Shifting Electoral Boundaries

Shifts in the electoral boundaries represent, for a population accustomed to constant spatial "adjustments", an additional and eloquent example of State defined territoriality.

These maps portray the results of selected parliamentary elections between 1963 and 2015, leaving out the elections of 1968, 1972, 1980, 1984, 1988, 1997, 2001 and 2011.

Between 1963, the year Singapore joined the federation of Malaysia, and 2015, when it celebrated its 50 years as an Independent Republic, 12 parliamentary elections have been held. In other words, over a 52-year period, a general election was held on average every four years. All were won by the People's Action Party (PAP), with its near total dominance was seriously challenged only a few times. The most significant challenge occurred in 1963, when it won only some 47 per cent of the vote but as many as 37 of the 51 constituencies, it main adversary, the Barisan Sosialis Party (BSP) having won 33 per cent but only 13 constituencies. Since then, the PAP's worst scores occurred in 1991, when it won only 61 per cent of the vote but 77 of the 81 parliamentary seats, and 2011 when it won barely 60 per

cent of the vote but 76 of the 82 parliamentary seats. Clearly, the electoral map has increasingly favoured the governing party. This even applies to the results of the September 2015 elections, when the PAP's dominance was somewhat re-established, when it won nearly 70 per cent of the vote and 83 of the 89 seats.

In the 1963 elections, the challenge from the BSP was manifest in the outlying areas, where agriculture was still largely practised. In 1968, however, when the first post-independence elections were held, the BSP boycotted them and voter turnout was very poor, with only 5 of the 58 seats challenged by the very weak opposition. The PAP won them all, 52 by walkover and, for the 6 seats that were contested, with 86 per cent of the vote. From then on, because of its own merit and a divided and small opposition (as many as nine parties were involved in 1976, 1984, 1988 and 2015), the governing party has won all national elections. In 1968, 1991 and 1997, it actually won more seats without any contest than through actual votes. Only in 1963, 1984 and 2015 did the opposition parties manage to field candidates in all constituencies.

Among other factors that may have prevented better electoral performances from the opposition parties, including the currently strongest one, the Worker's Party, two stand

Selected Election Results: By Constituency

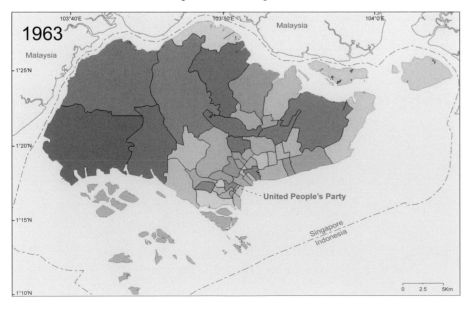

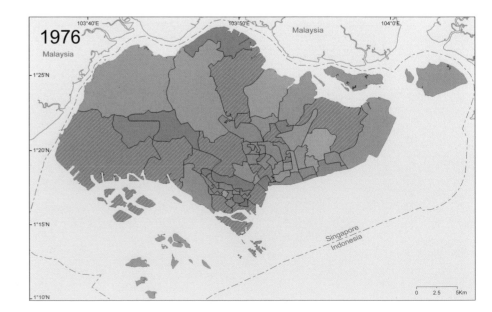

out. First, the establishment in 1988 of so-called Group Representation Constituencies, whereby citizens cast their vote not for a specific candidate but for a party constituency (represented by three to six candidates), which may carry as many as six seats and, second, changes in electoral boundaries. For, just like in many other countries, but perhaps more than in most, the Singapore electoral map has been modified repeatedly and sometimes drastically. Boundary changes are in fact the privilege of the cabinet and are usually only announced a few days before the elections. Whatever the impact on the actual electoral outcome of gerrymandering, shifts in the electoral boundaries represent, for a population accustomed to constant spatial "adjustments", an additional and eloquent example of state-defined territoriality.

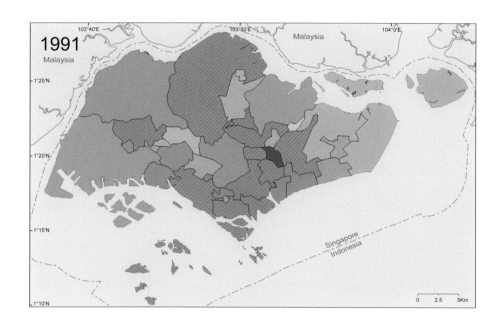

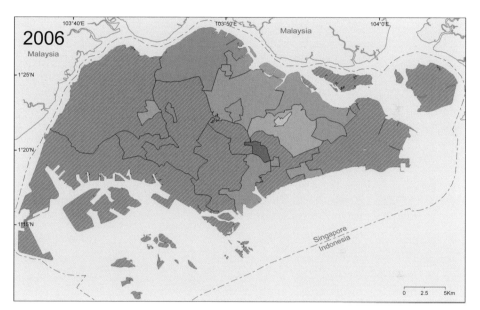

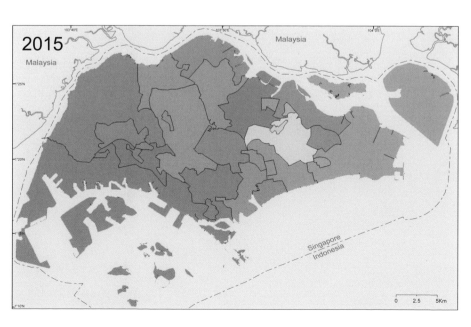

Total Votes Won (%)

	People's Action Party	United People's Party	Singapore Democratic Party
	Barisan Sosialis	Worker's Party	Walkover

Parties	1963		1976		1991		2006		2015	
	Constituencies	Votes	Constituencies	Votes	Constituencies	Votes	Constituencies	Votes	Constituencies	Votes
People's Action Party	37/51	272,924	69/69	590,168	77/81	477,760	82/84	748,130	83/89	1,579,183
Barisan Sosialis	13/46	193,301	0/6	25,411						
United People's Party	1/46	48,785								
Singapore Alliance	0/42	48,967								
Independents	0/16	6,788	0/2	4,173	0/7	14,596			0/2	2,780
Worker's Party	0/3	286	0/22	91,966	1/13	112,010	1/20	183,578	6/28	282,143
Partai Rakyat	0/3	8,259								
Pan-Malayan Islamic Party	0/2	1,545								
Parti Kesatuan	0/1	760								
Singapore United Front / United Front			0/14	53,373						
United People's Front			0/6	14,233						
Pertubuhan Kebangsaan Melayu Singapore			0/2	9,230	0/4	12,862				
Justice Party, Singapore			0/2	5,199	0/4	15,222				
People's Front			0/1	2,818						
Singapore Democratic Party					3/9	93,856	0/7	45,937	0/11	84,931
National Solidarity Party					0/8	57,306			0/12	79,826
Singapore Democratic Alliance							1/20	145,628	0/6	46,550
Reform Party									0/11	59,517
Singfirst									0/10	80,867
Singapore People's Party									0/8	49,107
People's Power Party									0/4	25,475
Total	51	581,615	69	796,571	81	783,612	84	1,123,273	89	2,290,379

Chapter 6

Going Global

"We inherited the island without its hinterland, a heart without a body."

– Lee Kuan Yew, 2000, p. 3

"We have neither a domestic market nor a hinterland. (…). To be competitive we must compensate for these disadvantages by offering more. (…). Our future lies in being plugged into the international network of trade and communications."

– Ministry of Trade and Industry, 1986, p. 12

One of Singapore's most impressive achievements over the last half century has been the globalisation of its economy. In searching for a new hinterland, it not only redefined its relations with its immediate neighbours but also launched itself into a veritable world expansion, where what is generally referred to as Singapore Inc has played an increasingly crucial role. Singapore Inc represents a unique form of state cum private sector partnership whereby the Singapore government is heavily involved in backing financially private entrepreneurs, particularly those going global. This is primarily the case of Temasek Holdings, a powerful sovereign fund owned by the Singapore government, which shares the risks as well as the profits of most of the large Singapore-based multinational enterprises. This has been manifest in nearly all fields of economic activity and well beyond the economic domain as even the education system has been developing increasingly broad international links. But Singapore did not start from scratch, its long-established status as a major sea port having provided it with good groundwork.

First, the island republic's links with the world were of a trading nature, its function as a sea and air transport hub being developed and exploited to the maximum as soon as it became independent. Second, closely related to the trading vocation, came the development of banking and financial services. As by the mid-1960s Singapore was turning itself into a frontier for industrial investments – thanks to the rapid and massive development of Jurong industrial estate – it began to develop a financial sector increasingly linked to international capital. Singapore was becoming a safe and reliable place to invest in

and do banking amidst a fast-developing Southeast Asia. Given its continuing practice of entrepôt trade, its own manufacturing exports and the local development of petroleum refining and petrochemical industries, merchandise trading grew to a point where it became one of the world's leading ports, and eventually the busiest!

Then the need came for the Island Republic to spread its wings and to expand regionally and globally. This meant increasing substantially its investments into the economy of its two immediate neighbours, Indonesia and Malaysia. These resulted in the Singapore-induced development of industrial platforms in the Riau Archipelago and in Johor state. With Singapore constantly upgrading its own manufacturing sector, more labour-intensive industries were relocated "next door". While Singaporeans' apparently insatiable need for leisure activities has been partially fulfilled by the development of playgrounds in both these areas, particularly in Indonesia, forms of investments have also evolved. More significantly, Singaporean investments, partly industrial, have moved further afield, with China as first recipient.

The City-State's increasing involvement with the world economy has continued to leverage its status and experience as, by now, the world's leading port and one of the leading trading centres, with Singaporean enterprises operating a number of terminals in several of the world's major ports. Finally, somewhat surprisingly, Singapore's worldwide presence also relies on the deployment of its armed forces, which are active in training activities as well as peace-keeping and humanitarian ones in numerous countries.

Paya Lebar Airport aerial view. Courtesy of National Archives of Singapore.

38 Banking on Singapore

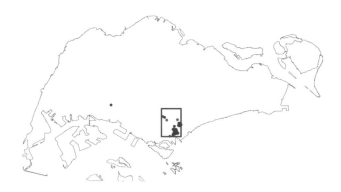

"Singapore has managed to transform itself dramatically to be more dynamic and to position itself as one of the best places to do business in the world."

– Gosh, 2015

Since its 1965 independence, Singapore has consolidated its position as a major transport hub, originally for maritime transport but increasingly for air transport. It has also become a leading business and financial centre, a hub in the world of money. It is not simply a banking centre but equally a haven for financial institutions of all types. It has in addition become one of the preferred headquarters for multinational companies in Asia, currently and increasingly at the expense of Hong Kong. The business world is banking not only in but also on Singapore, where, together, business, finance and insurance services accounted in 2015 for 28 per cent of GDP, versus 21 per cent for the manufacturing sector. The banking sector itself is therefore at the core of it all.

Singapore's local banks dominate the provision of services for consumers, with foreign players only cautiously allowed to increase their presence in this sector. After a series of consolidation moves, the local sector is represented by the Big Three: DBS Bank, the former Development Bank of Singapore, which also owns the POSB, the former Post Office Savings Bank; UOB, or the United Overseas Bank, which also includes Far Eastern Bank; and the Oversea Chinese Banking Corporation Ltd, or OCBC, including the Bank of Singapore. The Big Three now also have extensive operations in Asia.

For other banking activities, the field is open, though a bit difficult to track, as the Monetary Authority of Singapore (MAS) and the Association of Banks in Singapore (ABS) do not always use the same terminology to designate the various types of banks; the former having in fact modified some of its terminology over the years. Terms such as "full bank", "qualifying full bank", "wholesale bank", "merchant bank", "commercial bank", "offshore bank", "representative office of bank", "regional office of bank", are not always clear and often difficult to compare. At times, and unsurprisingly, the world of banking appears somewhat opaque.

Whatever the case, the world of capital is very visible in the Singaporean landscape, particularly and for obvious reasons in the Central Business District, also referred to as the financial district. Until recent years, banks and financial institutions remained largely concentrated south of the Singapore River and along or near Shenton Way and Maxwell Road. While this was true in 1975, by 2015, when the number of commercial and merchant banks had nearly doubled, the forest of bank towers bordering those two streets as well as Raffles Place and the Singapore River had spilled over into Marina Bay Financial Centre, with its gleaming blue towers. Just like the original financial core, this recently opened centre (2013) is not exclusively financial but also recreational, with the presence of high-end hotels and residences, restaurants and, of course, shopping malls.

Singapore's territorial gains at the expense of the sea are clearly reflected by the expansion of its banking and financial services sector and of its central role in Singapore's economy. This expansion also imprints itself quite spectacularly on Singapore's skyline.

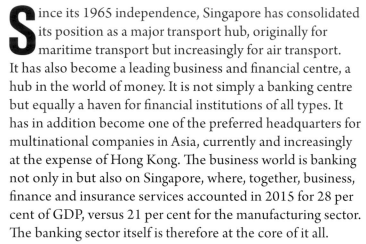

	1975	2015
Commercial banks	63	122
Merchant banks	20	35
Representative offices of banks	43	41
Regional offices of banks	21	
Total	147	198

Banks

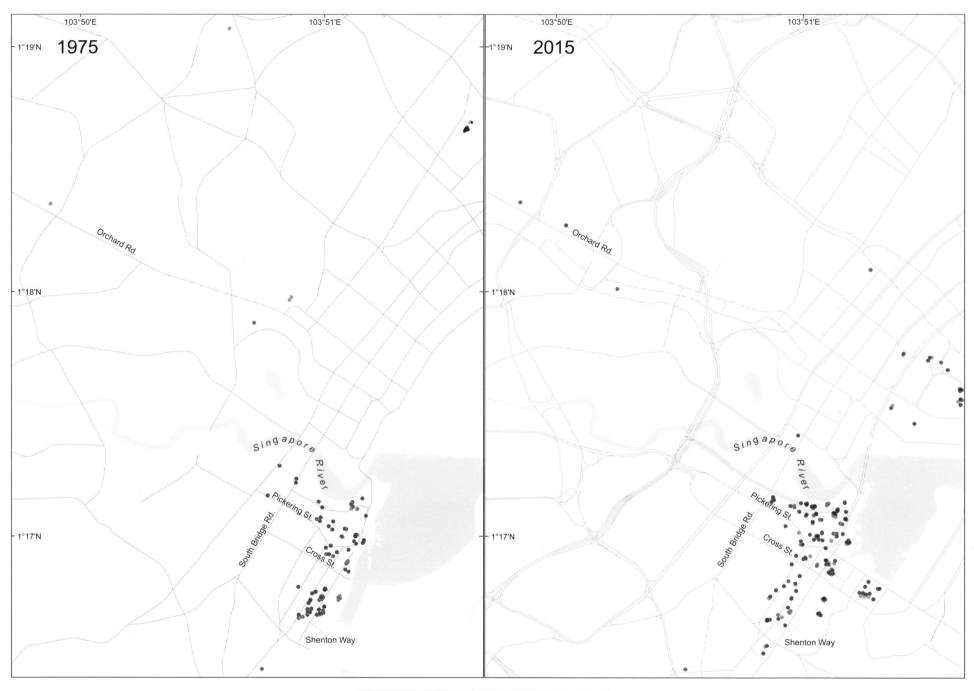

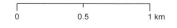

39

Doors Wide Open to the World

Singapore's doors on the world appear well open. But they are equally well guarded and that implies constant territorial transformations.

In 1960, Singapore's then unique seaport terminal, Keppel Harbour, handled 12 million tons of cargo. In 1989, the volume had reached 173 million tons, with petroleum accounting for nearly half of the total. That year, 39,000 ships visited its five terminals and 15 km of docks, all well serviced by 600 km² of port waters, the surface of the latter having doubled since 1960. This made Singapore the world's busiest port, and it has since maintained its rank as the world's leading maritime trade centre with, in 2015, some 133,000 ships stopping in its port waters and more than 575 million tons of cargo handled by its six terminals.

Nearly two-thirds of that cargo was containerised, and most of the bulk cargo was oil. On the south and southwest shore of the island, five terminals, are lined up – Tanjong Pagar, Brani, Keppel, Pasir Panjang (these four operated by PSA Singapore Terminals) and Jurong – with port facilities extending all the way to Tuas and into the islands lying offshore. The sixth terminal is on the northern shore of the island, next to the Sembawang shipyard, established in 1968 on the site of the former British naval base. The giant Brani terminal was opened in the 1990s on the island bearing the same name, just across from Keppel Harbour, which remains the centrepiece of the whole system. The network of terminals, primarily administered by Maritime and Port Authority of Singapore (MPA; formerly Port of Singapore Authority 1964–96), actually extends worldwide, given that PSA International manages an additional 37 terminals in 25 ports spread in 15 countries around the world (p. 122).

This represents another form of the expansion of Singapore, which, beyond its dynamism as a maritime trade centre, has developed an equally competitive air transport function. The island hosts two civil airports (Seletar and Changi) and four military air bases (Tengah, Sembawang, Paya Lebar and Changi), covering in total some 25 km², most

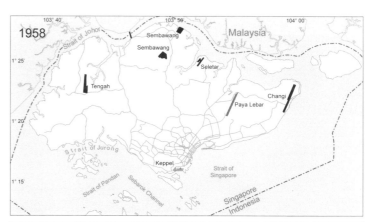

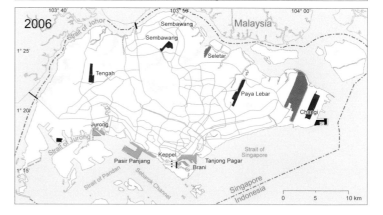

of this being the domain of the giant and still expanding Changi Airport. The military airbases as well as the naval bases have themselves gone through a number of relocations. For example, in the early 1970s, the Singapore navy's main base was on Brani Island. By the mid-1990s it was partially relocated on the Tuas peninsula. The Brani base was closed in 2000 and in 2004 a larger site was added off the Changi coast, just next to the Changi airport runways. In fact, civil and military functions do overlap, for example with the Singapore Air Force having an air base in Changi where it shares a runway with civil aviation. (p. 110)

Overall, military facilities occupy a substantial proportion of the national territory, particularly since the still expanding island of Tekong has been earmarked for exclusive military use (p. 24). Depending on how one defines military land, it occupies at least 10 per cent and possibly up to 20 per cent of the national territory (p. 130). Singapore's doors on the world appear well opened, even more since terminals specifically dedicated to cruise ships have been opened (p. 108). But they are equally well guarded and that implies constant territorial transformations.

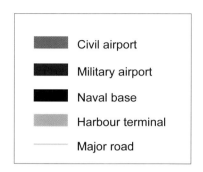

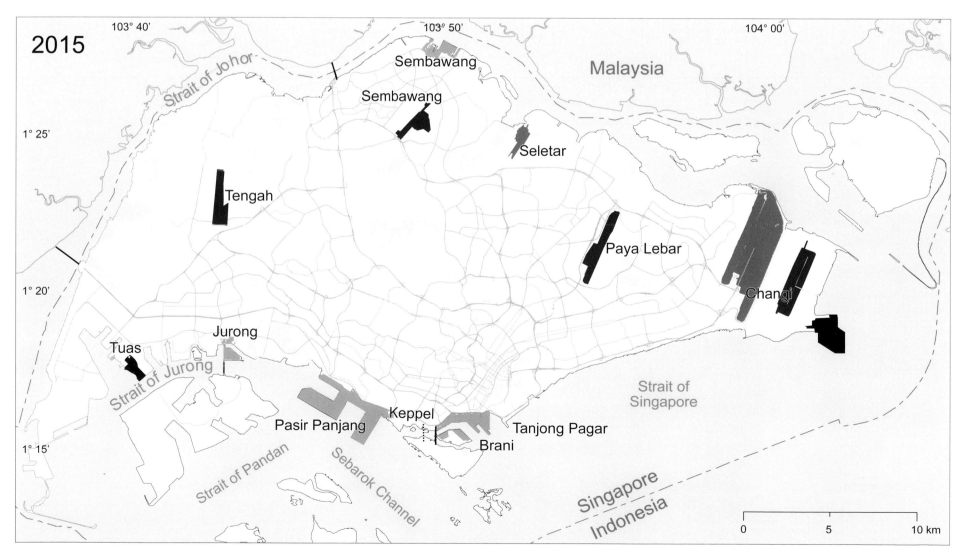

40

A Harbour City

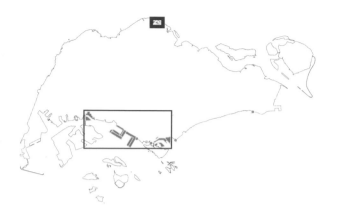

From inland, particularly from elevated viewpoints, such as downtown high-rise buildings, Mount Faber-Sentosa cable car and even portions of the coastal expressways, one can see, literally, mountains of containers and forests of cranes, the sight of which is even more impressive than that of the forest of banks lined up along the Singapore River.

Singapore's status as a harbour city has been an integral part of its history since well before it became a colonial outpost of the British empire in 1819. And since then, that role has taken on an increasingly global scale. It is therefore no surprise if during the first 50 years of the City-State's independence, its harbour functions have expanded, consolidated and diversified. This "Singapore imperative" implied territorial reallocation, transformation and expansion.

But, throughout these years of constant adaptation, the southern sea shore has remained the primary one opened to the world (p. 106). With the exception of the Sembawang wharves located on the Strait of Johor, just across from the rapidly expanding Malaysian city of Johor Bahru and on the site of the former British naval base – once touted as a key component of Fortress Singapore – all Singapore Harbour terminals are located on the southern flank of the of urban core. But these terminals have been spilling over towards the southwest shore, in the direction of the ever-expanding Jurong industrial frontier and seawards, as the gigantic and multifunction Pasir Panjang complex of five terminals illustrates. The creation and development of these terminals has been achieved largely through land reclamation, as testified by a comparison of the 2003 and 2015 satellite images.

While Jurong Island, in its capacity as petroleum hub, has been earmarked to handle Singapore's equally expanding trade in oil and gas (p. 68), bulk cargo trade is in the hands of the "continental" terminals. They are interspersed along some 20 km of the south and southwest sea shores, from Tanjong Pagar to Jurong, and have a total quay length of nearly 20 km. From inland, particularly from elevated viewpoints – such as downtown high-rise buildings, the Mount Faber–Sentosa cable car and even portions of the coastal expressways – one can see, literally, mountains of containers and forests of cranes, an even more impressive sight than the forest of banks along the Singapore River (p. 104).

Besides its function as one of the world's largest trading harbours, Singapore has recently been developing as a major cruise centre. To the passenger traffic handled by the smaller harbours lining up its southern shore and linking Singapore to some of its own peripheral islands and to the Riau Archipelago (p. 116), has been added a major international cruise centre. This has confirmed Singapore's increasingly evident intention to cater to the well-off regional and international tourists (p. 92).

Annual Cruise Ships and Passenger Throughout Volume, 1992–2014

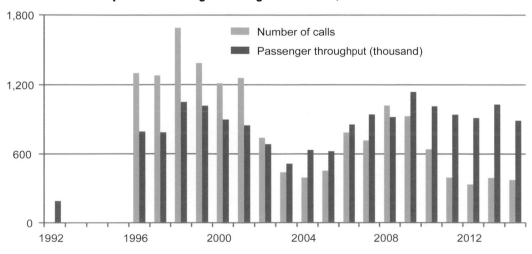

Facilities of Terminals, 2015

	Ro-Ro berths	Container berths	Quay length (m)	Area (ha)	Max depth at Chart datum (m)	Quay Cranes
Tanjong Pagar Terminal		7	2,100	80	14.8	27
Keppel Terminal		14	3,200	105	15.5	40
Brani Terminal		8	2,400	84	15	33
Sembawang Wharves		4	660	28	11.6	
Pasir Panjang Terminal 1		7	2,500	88	15	28
Pasir Panjang Terminal 2		7	2,300	120	16	28
Pasir Panjang Terminal 4		9	3,000	113	16	34
Pasir Panjang Terminal 5		5	1,850	110	18	22
Pasir Panjang Automobile Terminal 5	3		1,010	25	15	
Jurong port				155	15.7	

2003

Strait of Pandan

Container Cargo Throughputs, 1975-2014

TEUs (thousand)

35,000 — 28,000 — 21,000 — 14,000 — 7,000 — 0

1975 1980 1985 1990 1995 2000 2005 2010 2014

2015

Jurong Port

Strait of Jurong

Sembawang Wharves

PP Automobile Terminal

PP Terminal 1

PP Terminal 2

PP Terminal 3

PP Terminal 5

Future Terminals

Strait of Pandan

Keppel Distripark

Tanjong Pagar Terminal

Keppel Terminal

Brani Terminal

PP Pasir Panjang

Passenger Terminal or Cruise Centre

Limit of Port and Terminals

0   1   2 km

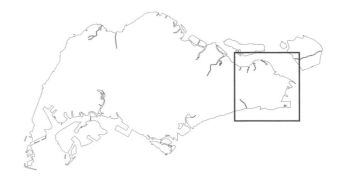

41 Changi: An Airport in the Sea

"Changi is a beautiful site at the eastern corner of the island […] The Airport and the pleasant 20-minute drive into the city made it an excellent introduction to Singapore, the best 1.5 billion investment we ever made."

— *Lee Kuan Yew, 2000, p. 204*

Like much of Singapore's industrial infrastructure, Changi airport, Southeast Asia's leading air transport centre, sits on reclaimed land. In 1958, when Britain's military installations were still operational, the island had two civil airports and ten military bases – occupying in total 10 per cent of the surface area of the island. Four of the military bases had runways. Following the 1968–71 British military withdrawal, the Singapore Armed Forces inherited part of the installations, including the Tengah air base, and the Seletar air base was handed over to the Civil Aviation Authority. Kallang airport was closed down, but Paya Lebar was maintained and eventually ceded to the military in July 1981, when passenger traffic, until then under its responsibility, was transferred to Changi. This huge airport had just been completed on the site of a former military runway, laid out in 1943 during the Japanese Occupation.

By 1981, that seaside airport had been thoroughly transformed and extended on reclaimed land. From then on its expansion has been steady. Changi airport now operates three runways, each 4 km long, and three full-blown terminals to which a terminal for budget airlines was added in 2006 but closed in 2012. In the meantime, the third "standard" terminal, of a size equivalent to that of the first two, was opened in 2008, sharing with the Singapore Air Force what was in fact the third runway, then already used by the SAF. So far, Changi has gained at least 13 square km on the sea. That expansion is slated to continue, as two additional terminals are planned, with the opening of Terminal 4 scheduled for 2017.

In the late 1950s and in fact well into the 1970s, this eastern tip of Singapore Island was primarily rural; the land

was mainly used for agriculture, and several Malay fishing villages dotted the shoreline. Now, as this shoreline continues to expand seaward, the airport surrounds itself, on the inland side, with vast industrial and trading installations, with ample land kept for parks and green areas. Perhaps even more than the Singapore harbour facilities, Changi

The Changi village area has been used for recreation since the colonial period, and its beaches and holiday bungalows were popular weekend retreats between the 1950s and 1970s.

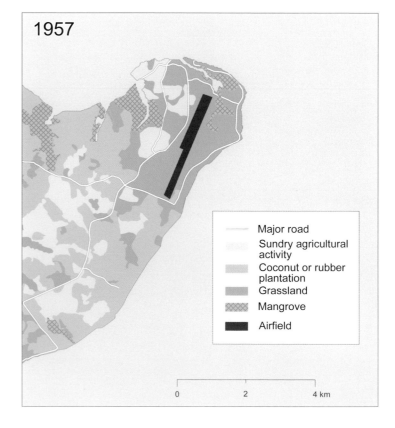

1957

	Major road
	Sundry agricultural activity
	Coconut or rubber plantation
	Grassland
	Mangrove
	Airfield

0   2   4 km

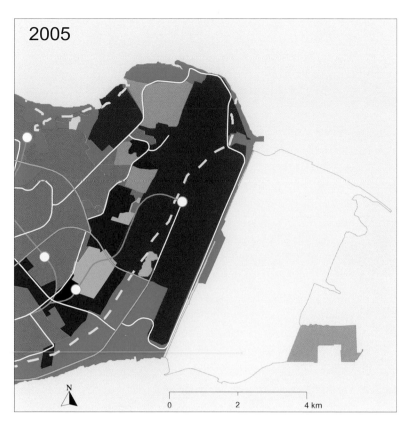

2005

Construction of Changi airport began in June 1975. It was officially opened in 1981.

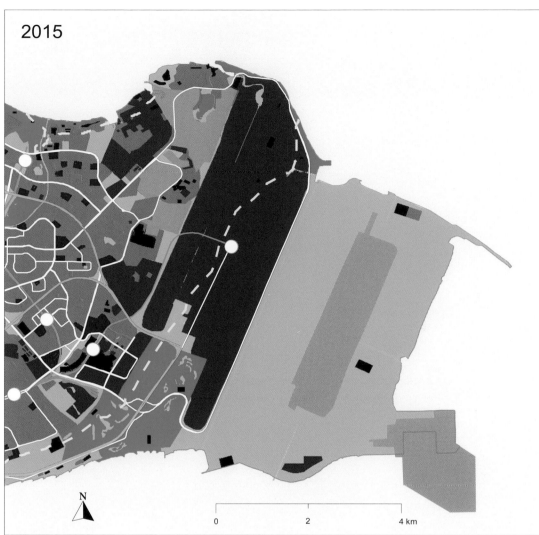

2015

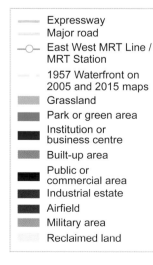

⎯⎯	Expressway
⎯⎯	Major road
⎯○⎯	East West MRT Line / MRT Station
⎯⎯	1957 Waterfront on 2005 and 2015 maps
▦	Grassland
▦	Park or green area
▦	Institution or business centre
▦	Built-up area
▦	Public or commercial area
▦	Industrial estate
▦	Airfield
▦	Military area
▦	Reclaimed land

airport – or Airtropolis according to Singapore government jargon – symbolises the small city-state's close ties to the world. In 2006, linked to 183 cities in 57 countries, it handled some 214,000 aircraft movements and more than 35 million passengers. In 2015, linked to 320 cities and 80 countries, it handled more than 346,000 commercial aircraft movements and some 55.4 million passengers flying with 100 airlines. In terms of international passenger traffic, this made Changi the world's seventh busiest airport and the second busiest in Asia. And probably more significant, it remains, arguably, the most pleasantly efficient international airport in the world. Since 1981, it has won over 500 awards, including 28 "Best Airport" awards in 2015 alone.

As part of its expansion, Changi has plans for three future terminals. Terminal 4, built on the site of the previous Budget Terminal, will be ready in 2017, and Terminal 5 will be open by the mid-2020s.

42 SIA: A World Class Airline and its Siblings

"In July 1972, I spelled out the need for a Singapore airline to be competitive and self-supporting: it would close down if it incurred losses."

– Lee Kuan Yew, 2000, p. 202

In 1965, when Singapore separated from Malaysia, the two countries continued to operate Malaysia-Singapore Airlines jointly. When this joint partnership was terminated in 1972, two airlines were formed: Singapore Airlines (SIA) and Malaysian Airlines System (MAS), the "system" having since been dropped. Both have become quite successful, SIA in particular, though MAS recently suffered considerably from a few dramatic setbacks. In 2005, the SIA network, together with that of Silk Air – its wholly owned subsidiary, which began operating under that name in 1992 and which limits its services to the Asian regional market – extended to 84 destinations in some 30 countries. Ten years later, their combined network had extended to 107 destinations in 36 countries.

Over recent years, several new routes have been opened by SIA towards Europe in one direction and Australia and New Zealand in the other. Meanwhile, Silk Air, in its capacity as SIA's premier regional wing, has increased substantially its coverage of cities located within Southeast Asia, the Indian subcontinent and Southern China.

SIA is largely recognised as one of the world's best airlines and also one of the most financially successful. It benefits from the fact that it does not need, as do many other airlines, to service isolated and less lucrative national markets. It is strictly an international airline, its network reflecting to a large extent the global reach of Singapore Inc. The latter is largely present in industrial countries, but has yet to make a breakthrough in Latin America and in much of Africa. Other SIA big assets are the quality and relative youth as well as frequent renewal of its aircraft fleet, and its constant opening of new routes. For example, in 2004, it launched non-stop flights from Singapore to Los Angeles and to Newark, near New York, in both cases a world's first. But these routes were abandoned in 2013 and replaced by more practical and lucrative ones via Tokyo and Frankfurt respectively. Over recent years, several new routes have been opened by SIA towards Europe in one direction and Australia and New Zealand in the other. Meanwhile, Silk Air, in its capacity as SIA's premier regional wing, has increased substantially its coverage of cities located within Southeast Asia, the Indian subcontinent and Southern China.

Although SIA has been privatised, the Singapore government still holds a majority of its shares. And, as so many Government-Linked Companies, it is constantly expanding, by diversifying its financial assets, including through the purchase of shares in other airlines, such as Virgin and Indian Airlines or two other airlines based in Singapore. These are Tiger Airways, better known as Tigerair, and Scoot, both low-cost carriers. Since it began operations in 2004, Tigerair has grown very rapidly, serving in 2015 nearly 40 destinations, all in the Asian region and within a five-hour flying time radius. In 2012, Scoot began operating out of Changi flies to cities such as Tokyo and Melbourne which figure among its predominantly East Asian and Australian destinations. It is scheduled to open service to Jeddah (its 17th destination), in late 2016, by which time its entire fleet will be made up of Boeing 787, the Dreamliner aircraft.

The four airlines, operating out of Changi airport and all owned at least partially by SIA, itself more than half-owned by the Singapore government, are continuously extending their reach. By 31 March 2016, the combined networks of the island republic flag carriers were operating some 170 aircrafts and serving nearly 170 cities. The overall coverage clearly emphasised Singapore's Asia-focused globalisation process (pp. 114–22).

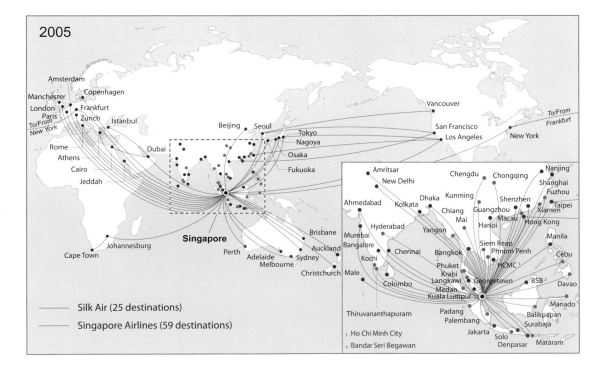

2005

Silk Air (25 destinations)
Singapore Airlines (59 destinations)

1. Ho Chi Minh City
2. Bandar Seri Begawan

Route Map for Singapore Airlines and Subsidiaries

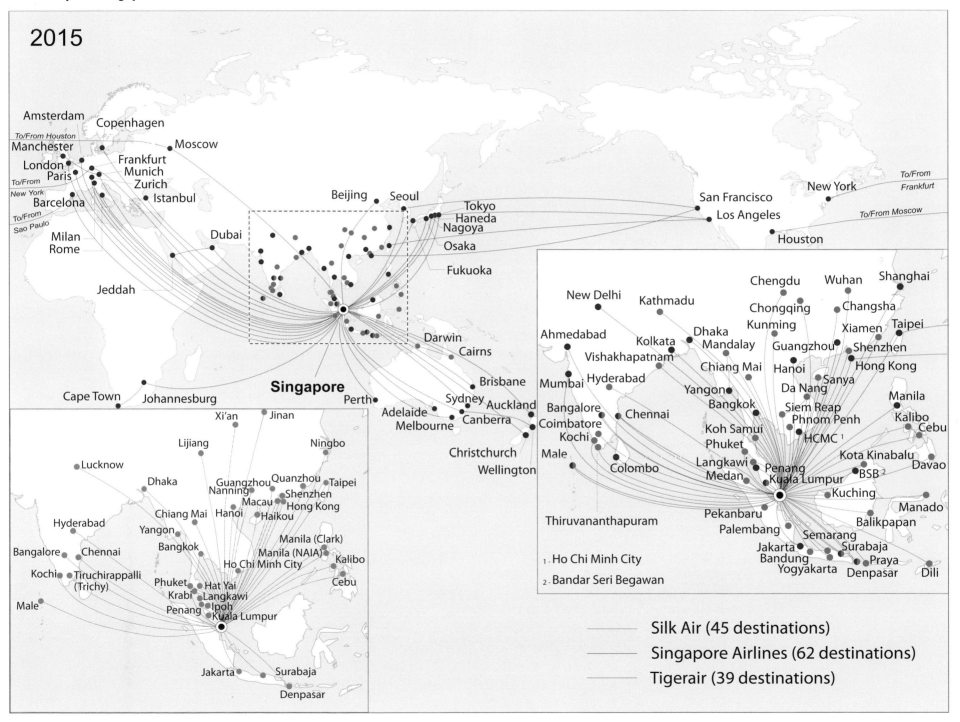

2015

Amsterdam
Copenhagen
To/From Houston
Manchester
London
Paris
To/From New York
Barcelona
To/From Sao Paulo
Frankfurt
Munich
Zurich
Istanbul
Milan
Rome
Dubai
Jeddah
Moscow

Beijing
Seoul
Tokyo
Haneda
Nagoya
Osaka
Fukuoka

San Francisco
Los Angeles
Houston
New York
To/From Frankfurt
To/From Moscow

Darwin
Cairns
Brisbane
Sydney
Auckland
Adelaide
Melbourne
Canberra
Christchurch
Wellington

Singapore
Perth

Cape Town
Johannesburg

New Delhi
Kathmadu
Ahmedabad
Kolkata
Dhaka
Mandalay
Chengdu
Wuhan
Shanghai
Chongqing
Changsha
Kunming
Xiamen
Taipei
Guangzhou
Shenzhen
Vishakhapatnam
Chiang Mai
Hanoi
Hong Kong
Mumbai
Hyderabad
Yangon
Bangkok
Da Nang
Sanya
Bangalore
Chennai
Siem Reap
Phnom Penh
Manila
Kalibo
Cebu
Coimbatore
Koh Samui
HCMC ¹
Kochi
Phuket
Kota Kinabalu
Male
Langkawi
Penang
BSB ²
Davao
Medan
Kuala Lumpur
Kuching
Manado
Thiruvananthapuram
Pekanbaru
Balikpapan
Palembang
Semarang
Jakarta
Surabaja
Praya
Bandung
Yogyakarta
Denpasar
Dili

1. Ho Chi Minh City
2. Bandar Seri Begawan

Xi'an
Jinan
Lijiang
Ningbo
Lucknow
Dhaka
Guangzhou
Quanzhou
Nanning
Taipei
Macau
Shenzhen
Chiang Mai
Hanoi
Hong Kong
Hyderabad
Yangon
Haikou
Bangalore
Chennai
Bangkok
Manila (Clark)
Kochi
Tiruchirappalli
(Trichy)
Manila (NAIA)
Ho Chi Minh City
Kalibo
Cebu
Male
Phuket
Hat Yai
Krabi
Langkawi
Penang
Ipoh
Kuala Lumpur
Jakarta
Surabaja
Denpasar

——— Silk Air (45 destinations)
——— Singapore Airlines (62 destinations)
——— Tigerair (39 destinations)

43 Trading with the World

"By the late 1970s […] we had found our new hinterland in America, Europe and Japan."

– *Lee Kuan Yew, 2000, p. 63*

When the Association of Southeast Asian Nations (ASEAN) was founded in 1967, one of its stated objectives was to encourage regional economic integration, notably through trade. During the preceding decades, trade between the region's countries had been losing ground to extra regional trade. During the 1970s and 1980s, and contrary to declarations from several of the region's political leaders, the trend was not reversed, the proportional share of intraregional trade having in fact declined. That applied to overall ASEAN trade as well as to Singapore's own international trade. But by the 1990s, as its global trade kept growing, the share of that trade with ASEAN neighbours did begin to increase significantly, representing a third of Singapore's exports in 1995 and a third of its imports by 2000.

However, over the last 15 years, the pursuit of the 1967 objective has resumed: the share of Singaporean imports from ASEAN countries did drop substantially, down to 20 per cent, while the share of its exports towards its neighbours, after an initial decrease, has rebounded over the last decade, up to 28 per cent. Nevertheless, the strong and continuous growth of Singapore's external trade has gradually required a market much larger than that of ASEAN countries alone. Among the latter, Malaysia remains the number one partner, twice as important as the second, Indonesia.

Over the last quarter century, strongly emerging partners have been, unsurprisingly, South Korea and China, by now the leading global partner, way ahead of Japan, which had remained dominant for several decades. As a result, Singapore has seen East Asia's share of its exports nearly double over the last quarter century, but with only a slight increase in the share of its imports. Meanwhile the share of both the USA and the European Union has been reduced notably, particularly as a destination for Singaporean exports.

What appears possibly more significant is that overall commercial integration within ASEAN is increasing, at least

Billions of $US	Imports		Exports	
	1990	2014	1990	2014
ASEAN countries	12.5	73.4	13.7	114.5
Brunei	0.1	0.2	0.5	1.9
Cambodia	0.0	0.3	0.2	1.1
Indonesia	1.9	16.8	1.3	25.2
Laos	0.0	0.0	0.0	0.1
Malaysia	8.3	39.0	6.9	49.0
Myanmar	0.1	0.2	0.2	2.4
Philippines	0.3	5.0	0.7	6.9
Thailand	1.6	8.8	3.5	15.0
Vietnam	0.1	3.2	0.4	12.9
Rest of the World, including:	48.3	292.8	39.0	295.2
Australasia	1.3	5.6	1.5	17.7
Australia	1.2	4.7	1.3	15.5
New Zealand	0.1	0.9	0.2	2.2
East Asia (three countries)	16.1	86.1	6.6	84.9
China	2.1	44.4	0.8	51.5
Japan	12.3	20.1	4.6	16.7
Rep. of Korea	1.8	21.6	1.2	16.7
European Union	8.2	43.8	7.8	32.9
North America	10.1	39.1	11.7	25.2
Canada	0.4	1.2	0.5	1.0
USA	9.7	37.9	11.2	24.2
WORLD	60.8	366.2	52.7	409.8
Share of ASEAN countries	21%	20%	26%	28%
Share of Australasia	2%	2%	3%	4%
Share of East Asia	27%	24%	12%	21%
Share of EU	14%	12%	15%	8%
Share of North America	17%	11%	22%	6%
Share of all others	21%	32%	22%	33%

for Singapore. Over and beyond fluctuations, Singapore's exports to the ten regional partners have increased in relative terms. While its global trade has expanded massively, and with a growing number of partners, it has gradually become more

Asia focused, thanks to ASEAN and ASEAN 2, the latter confirming the increasing role of Australia and New Zealand among Singapore's economic partners.

Singapore imports and exports from various countries and regions (in billions of USD), in 1990 and 2014

Singapore's Trade with ASEAN Compared to its Total Trade

ASEAN countries   Rest of the World   Share of ASEAN countries

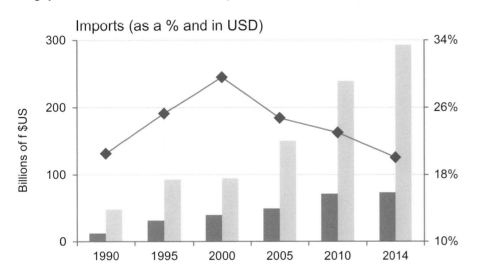

Imports (as a % and in USD)

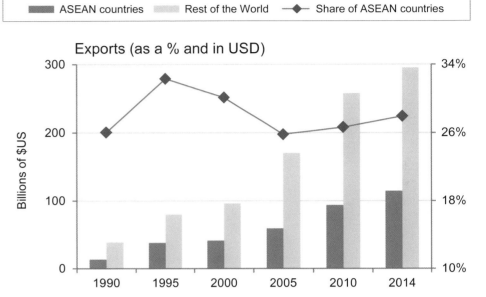

Exports (as a % and in USD)

ASEAN
Autralasia
East Asia
European Union
North America

0   4,000 km

44 Foreign Lands for Expansion: The Riau Islands

In the case of Indonesia, the primary target area for Singaporean industrial expansion
has been some of the neighbouring Indonesian islands of the Riau Archipelago, which
extends between Singapore and the large island of Sumatra.

The requirements of industrial expansion – along with thirst for water and hunger for soil and sand to expand Singapore's land surface – have led to a series of deals with Malaysia and Indonesia (pp. 23 and 28). But collaboration has gone way beyond that and to the advantage of the island republic's industrial expansion, in need of cheaper labour (pp. 116 and 118).

In the case of Indonesia, the primary target area has been some of the neighbouring Indonesian islands of the Riau Archipelago, which extends between Singapore and the large island of Sumatra. For a time, some of the smaller islands remained primarily a source of building material, but a few of these and particularly two among the larger ones (Batam and Bintan) gradually became targets for the City-State's investments in manufacturing and for the establishment of playgrounds for Singaporean as well as other foreign tourists.

Riau Islands Province, created in 2002, is actually composed of three major groups of islands, whose total population reached barely two million people in 2015. The first group, which corresponds to the Riau Archipelago proper, extends westwards from the southern entrance of the Malacca Straits and includes, from west to east, islands such as Karimun Besar, Kundur, Batam, Rempang and Bintan. The second group is located much further south and centres on the Lingga Archipelago. The third is situated far to the northeast, well into the South China Sea, and comprises the Anamba and Natuna Islands (p. 10).

It is the Riau Islands, or Riau Archipelago proper, which are part of what has been known since the 1980s as the Singapore/Johor/Riau (SIJORI) growth triangle. This was officially recognised by the three countries in 1994. Situated south and southeast of Singapore, these islands were heavily targeted by the overflow of Singaporean

industrial capital. Batam is by far the most populated in the province, containing well over half of its total population. The 715 km square island is located 20 km southeast of Singapore to which it is approximately equivalent in size: it plays the role of an offshore industrial platform, a site offering relatively inexpensive space and labour. Singapore-based entrepreneurs have transferred some of their activities onto the island and invested in several ventures. In typical Jurong fashion, land reclamation was carried out on the northern coast of Batam, largely at the expense of mangrove forest and coral reefs.

This industrial frontier, located on Indonesian territory and relying on Indonesian labour, has spread into other islands, including Bintan. On the southern side of Bintan, the huge Batamindo Industrial Estate was by 2007 already employing some 70,000 workers, most of them women. Until then migration into these islands had been such that the growth of the province's population was for a time the fastest in Indonesia.

Singapore's investments also reach into the recreational sector with the multiplication of golf courses and beach resorts, including the very large Bintan International Tourist Resort on the northern shore of the island. Both Batam and Bintan Islands are within easy reach for Singaporean entrepreneurs and tourists, with over 25 ferry trips per day from Singapore, with the majority heading for Batam's five harbours and transporting primarily Singaporean vacationers.

The appeal of the Riau islands for Singapore investors, still the leading ones, originally stemmed from industry, including heavy industry – such as the string of shipyards on the western coast of Batam employing more than 3000 workers. However, with the recent decline in the global economy, those investments have been stalling. Indonesian migrant labourers, who had moved by the hundreds of thousands to Batam and Bintan during the boom years in the 1980s to mid-2000s, have been finding work conditions more difficult. Retrenchments have begun and so has labour unrest. Singapore's industrial frontier is not so compliant anymore. However, while industrial expansion into the Riau islands has been slowing down, tourism has apparently continued to grow, primarily thanks to tourists from Singapore. The islands targeted are not only Batam and Bintan but also several others, including a string of six islands extending south of Batam and linked by bridges. For the moment, the Riau Archipelago remains very much part of Singapore's hinterland.

Well-known as part of Singapore's recreational hinterland, the Riau islands of Batam and Bintan also belong to its industrial frontier.

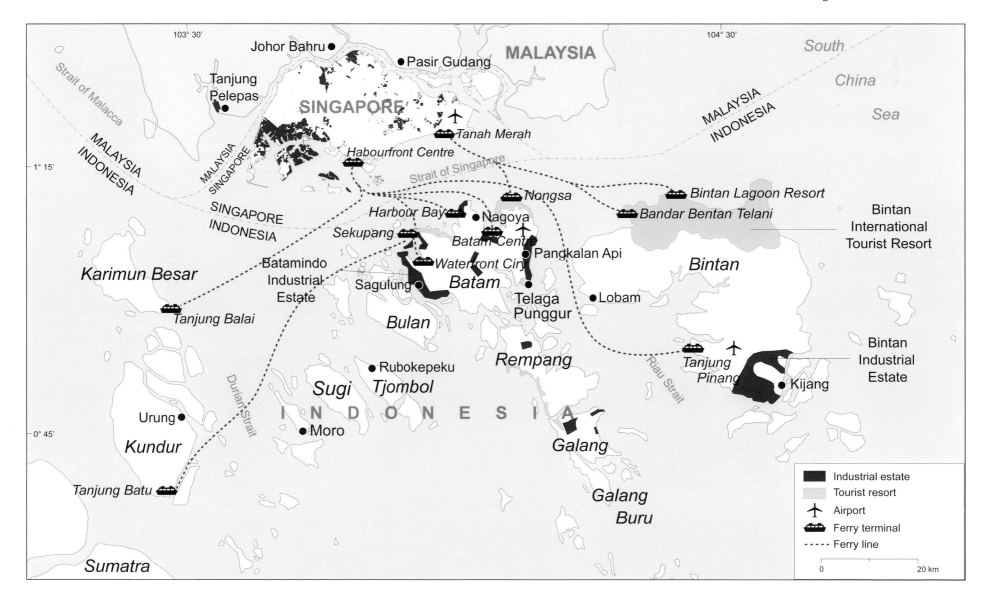

45 Foreign Lands for Expansion: Johor

"This deal with Temasek/CapitaLand signifies the entry of a big Singaporean investor that sees potential in Iskandar Malaysia, the southern economic corridor of Malaysia, and could signal entry of more Singaporean investors in Iskandar Malaysia"

– The Star, 19 December 2013

Singapore Between Lands for Expansion

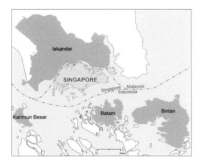

Singapore's size in relation to Iskandar (Johor, Malaysia), Batam, Bintan and Karimun Besar (Indonesia).

In the case of Malaysia, the southern part of Johor state has also since the early 1980s become the site of a number of industrial development ventures, largely financed by Singaporean capital. Officially regrouped in 2006 into the South Johor Development Region, perhaps better known as the Iskandar Development Region (IDR), this industrial frontier encompasses some 2,200 km², an area three times the size of Singapore.

According to the Ninth Malaysia plan (2006–10), it was expected that by 2010 some 1,000 factories and businesses, many linked to the information technology sector, would be employing more than 120,000 workers. From the Malaysian point of view, this was to limit the congestion that had been building up around Kuala Lumpur and the so-called Multimedia Super Corridor situated south of it. It is not clear whether the 2010 employment target was reached. However, according to declarations made by a Johor minister in June 2015, more than 650,000 jobs were created within the project between 2007 and 2014, primarily in manufacturing, hospitality, food and beverage and education industries.

The capital city of Johor state, located just across the causeway from Singapore, has become the command centre of this new growth region. Johor Bahru, or JB as it is known familiarly, has itself been growing very rapidly, the population of its metropolitan region having risen between 1991 and 2015 from some 450,000 inhabitants to nearly two million inhabitants, making it Malaysia's second largest conurbation after Kuala Lumpur. This is attributable to its role both as capital of Johor State and of the IDR, but also as receiving centre for a number of additional Singaporean activities, which include recreation, shopping, investment in housing, etc.

Apart from a number of industrial parks expanding all over the interior of the Iskandar region, three major harbour centres are fast developing on the shores of Johor Strait. Located at its western extremity, the most important of the three is Tanjung Pelapas harbour. Backed by its own industrial park, it is already competing with Singapore's Tanjong Pagar terminal for container handling, while the two other Malaysian ports on Johor Strait are also expanding rapidly. These are Pasir Gudang, also called Johor port – which handles shipbuilding, petrochemicals as well as oil palm storage and redistribution – and Tanjung Langsat – which specialises in petrochemicals and liquid bulk handling. Both ports also maintain large oil storage facilities.

Among the assets of the Iskandar project and the State of Johor are the fact that the latter can associate major agricultural functions, particularly oil palm cultivation and processing, with other industrial and commercial functions, largely financed by foreign capital. Among the top five foreign investors, which between 2007 and 2014 committed over 30 billion ringgit (about 10 billion SGD), Singapore is still the most important one, responsible for nearly 40 per cent of the total.

Since 2006, by which time the major industrial infrastructure had been laid out, development planning has centred on five so-called flagship zones. These form a broad triangle, at the extremities of which are located, the "western development gate", centred on the port of Tanjung Pelepas with its free trade zone and industrial hinterland; the "eastern development gate", centred on the port of Pasir Gudang with its own manufacturing components; and at the top the Senai-Skudai airport city and development hub, characterised by manufacturing in electronics and petrochemicals as well as

Flagship Zones

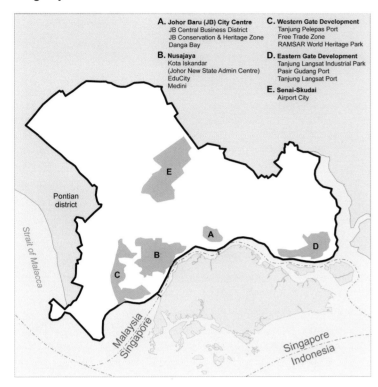

A. Johor Baru (JB) City Centre
JB Central Business District
JB Conservation & Heritage Zone
Danga Bay

B. Nusajaya
Kota Iskandar
(Johor New State Admin Centre)
EduCity
Medini

C. Western Gate Development
Tanjung Pelepas Port
Free Trade Zone
RAMSAR World Heritage Park

D. Eastern Gate Development
Tanjung Langsat Industrial Park
Pasir Gudang Port
Tanjung Langsat Port

E. Senai-Skudai
Airport City

The Iskandar Development Region
surrounds Singapore island, which
it dwarfs with its 2200 km² versus
the City-State's 720 km².

Iskandar Development Region Flagship Zones

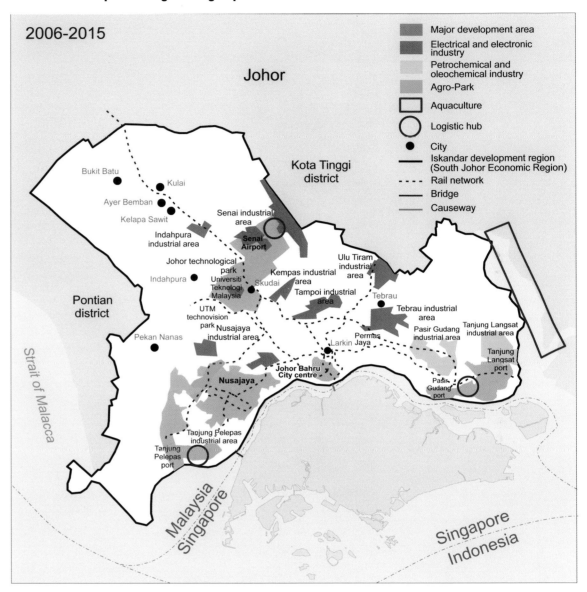

Iskandar development region
flagship zone.

harbouring a 'Cybercity'. The other two flagship zones, also
called "key economic driving areas", are also located near
Johor Strait, hence just across from Singapore. These are,
first, the city of Johor Bahru, the financial, cultural and tourist
centre of the state; and, second, Nusajaya (Iskandar Puteri aka
Nusajaya), a large urban industrial and property development
area centred on Kota Iskandar, the new administrative centre
of Johor State.

And Singaporean investors, with CapitaLand at the
forefront, are very much active in urban development,
particularly on the southwestern flank of Johor Bahru, notably
in the Danga Bay area, where huge investments are at stake.
This is occurring in a context where the urban expansion of
Nusajaya involves land reclamation by Malaysia into the waters

of Johor Strait. There is also the case of the so-called Forest
City, a 20 km² Chinese-funded high-rise housing development
on reclaimed land near the port of Tanjung Pelepas. This
Singapore-type expansion on the part of the city-state's
neighbour seems to trouble the authorities. Could it be that
the island republic is being challenged at its own game by its
Malaysian partner?

46 Foreign Lands for Expansion: The World

Singaporean investments abroad confirm the City-State's ambitions to develop close economic links with the two economic superpowers, China in particular.

Singapore's economic expansion has of course gone well beyond the shores of immediate neighbouring lands. Firstly, besides the remarkable reach of its "ambassador" airline and its siblings (p. 112), Singapore manages a rather impressive network of diplomatic missions – impressive at least for a country of its size – as it currently maintains Singaporean Embassies and High Commissions in 28 countries, as well as consulates in an additional 26. Consulates are also maintained in 10 of these 28 countries hosting Embassies or High Commissions. The most favoured, given their overall weight in world affairs, each hosting four additional consulates, are China and the USA.

Secondly, the City-State's global reach manifests itself in a number of additional ways, particularly in the economic field. For example Singapore Telecom, better known as SingTel, already the largest company by market capitalisation listed on the Singapore Exchange, has been expanding very aggressively in recent years, particularly within Asia. Majority owned by Temasek Holdings, the investment arm of the Singapore government, it was already operating in 17 countries in 2005. It now has offices in 46 cities located in 21 countries and employs more than 23,000 people

Numerous other government-linked companies (GLC), as they are known locally, can also count on the government to help them expand globally. This can be achieved through International Enterprise (IE) Singapore, an offshoot of the former Singapore Trade Development Board. This can be achieved through International Enterprise (IE) Singapore, the upgraded version of Singapore's old Trade Development Board. Operating under Singapore's Trade and Ministry Industry, it maintains a network of 39 representative offices throughout the world, which helps link Singapore enterprises – the so-called Singapore, Inc – with new markets and international partners. The Economic Development Board

(with 22 overseas offices) advises companies who want to invest in Singapore. In 2005, IE Singapore had offices in 36 cities distributed among 22 countries, with eight of these cities located in China. By 2015, the numbers had risen respectively to 40, 25 and 11. The presence of so many offices in China alone reveals the extent to which Singapore has been developing its economic links with the Asian superpower.

Singaporean investments are themselves dispersed throughout the world with a major focus on Asia, Asian countries hosting more than half of the total. Between 2005 and 2014, such investments more than tripled, from 200 to 620 billion SGD. Throughout the decade, China remained the primary recipient of Singaporean investment, receiving nearly 110 billion in 2014! Other major recipients were, in Asia, Hong Kong, Malaysia and Indonesia, with Australia the first among non-Asian countries. However that does not take into account two well-known tax havens, the Cayman Islands and the British Virgin Islands, which, together, took in nearly as much Singaporean "investment" as China!

Singaporean enterprises also include major real estate actors such as CapitaLand which, as it claims on its own website, has "interests in and manages a pan-Asian portfolio of 104 shopping malls across 54 cities in Singapore, China, Malaysia, Japan and India, with a total property value of 40.4 billion SGD[1]." Closely related are enterprises such as Surbana International Consultants, a leading consultancy specialising, in their own words, in "sustainable urban solutions" [2]. Headquartered in Singapore and jointly owned by Temasek Holdings and CapitaLand, it has expanded its consultancy business to over 90 cities in 26 countries, also predominantly

---

[1] http://investor.capitaland.com/misc/ar2015/files/assets/basic-html/page77.html

[2] http://www.surbana.com/profile/

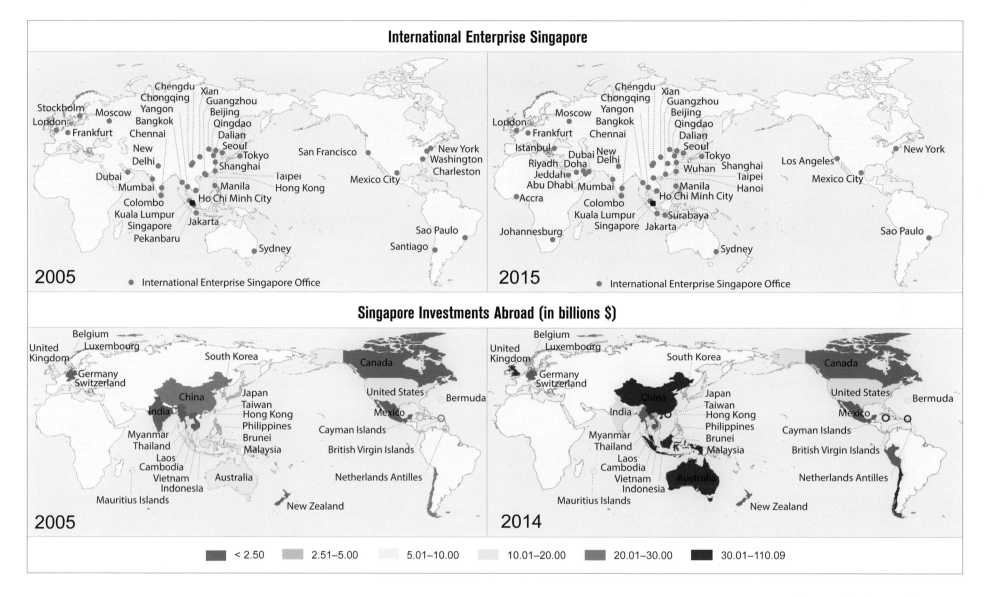

International Enterprise Singapore

2005 — International Enterprise Singapore Office

2015 — International Enterprise Singapore Office

Singapore Investments Abroad (in billions $)

2005

2014

| < 2.50 | 2.51–5.00 | 5.01–10.00 | 10.01–20.00 | 20.01–30.00 | 30.01–110.09 |

Asian. Finally, Jurong Consultants must be counted among major Singaporean development advisers; given the success of the Jurong Industrial Estate in Singapore itself, it has become a leading consultant for foreign countries intent on developing their own industrial estates. Leading clients include China, India and Abu Dhabi.

Singapore has increased its investments in almost all regions of the world from 2005 to 2014. Currently, its largest investments are in China, Malaysia, Indonesia, Australia and the United Kingdom. Investments in Chile, while significant in 2005, had dwindled by 2014.

47 The World as a Harbour

Besides PSA International, several other Singaporean enterprises have been acting as if the world was their harbour. Such is the case with Keppel Corporation (KC), which specialises in offshore and marine activities, property and infrastructure businesses.

Singapore's commercial expansion, as illustrated by the expansion of its management of terminals in foreign harbours, is predominantly concentrated in Eurasia, with limited involvement in the Americas.

Not only is Singapore the busiest port in the world (pp. 106 and 108), the port of Singapore also counts as the world's leading maritime trade centre. Not only does PSA Singapore operate most of the harbours and terminals that constitute the port of Singapore, PSA International Private Ltd, its international arm, also operates a number of overseas terminals located in foreign countries. In fact, PSA International comprises PSA Singapore terminals, PSA HNN and PSA Marine.

In 2005, PSA International was operating 28 terminals located in 26 foreign ports, in 15 countries. By 2015, the number of terminals had reached 34, distributed among 24 ports, still in 15 countries. Between those two years, PSA International's direct global presence has increased, although not in the number of countries and port where it is involved, but in the number of terminals it operates and more importantly in the size of these terminals, for example in the large ports of Kolkata and Antwerp, Europe's second largest after Rotterdam.

Despite its presence in the Middle East, Europe and South America, PSA has remained primarily involved in Asia, particularly in China and India where it operates respectively ten and five terminals, with a total designed capacity of over 18,000 and 15,000 20-foot equivalent units (TEUs). The 2015 combined designed capacity of foreign terminals officially operated by PSA reached nearly 57,000 TEUs. For that same year, the total capacity of the four container terminals operated by PSA Singapore Terminals (i.e., Brani, Keppel, Tanjong Pagar and Pasir Panjang, Jurong being operated independently) was about 35,000 TEUs. This figure illustrates eloquently the port of Singapore's relative importance in the PSA "empire". In fact, according to PSA International's 2015 Annual Report, nearly half of its revenue was generated by its four terminals in Singapore.

PSA International's global reach involves much more than its direct management of terminals in at least 15 countries; it also has partnerships with or acquisitions of rival companies. For example, in 2006 PSA International acquired a 20 per cent stake (4.4 billion USD) in Hong Kong's Hutchison Port Holdings with whom it formed a global partnership. Since then, it has continued its acquisitions and expansion, to the point of becoming the world's largest port operator.

Besides PSA International, several other Singaporean enterprises have been acting as if the world was their harbour. Such is the case with Keppel Corporation (KC), which specialises in offshore and marine activities, property and infrastructure businesses. Like so many Singaporean enterprises, KC owes its origin to Temasek Holdings which founded Keppel Shipyard when Keppel Harbour was taken over from the British Royal Navy after its withdrawal from the island in 1968. Today KC is composed of several divisions, one of the most successful being Keppel Offshore & Marine. Keppel O&M has become a global leader not only in offshore rig design, construction and repair, but also in ship building.

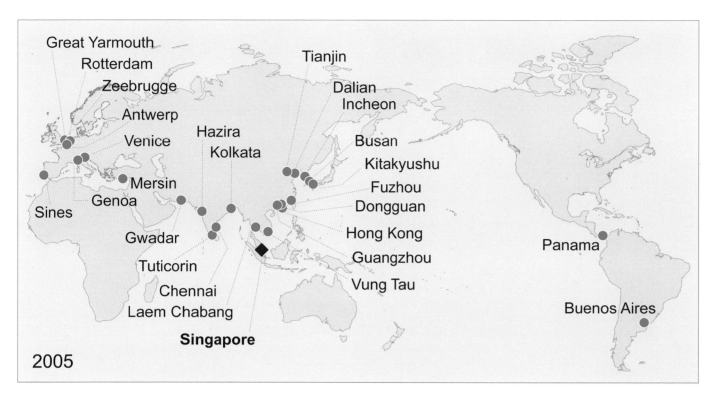

2005

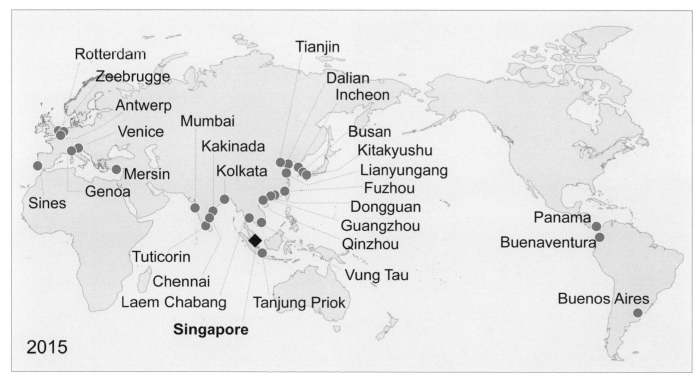

2015

Foreign Terminals operated by PSA International in 2015
Designed capacity (000 TEUs)

Thailand: (1 port, 1 terminal) Laem Chabang	2,200
Vietnam: (1 port, 1 terminal) Vung Tau	2,200
Indonesia: (1 port, 1 terminal) Tanjung Priok	1,500
China (7 ports, 10 terminals) Dalian (2) Tianjin Lianyungang Fuzhou (3) Guangzhou Dongguan Qinzhou (Beibu Gulf)	5,000 1,850 2,800 3,550 1,100 1,000 3,000
South Korea (2 ports, 2 terminals) Incheon Pusan	1,500 2,200
Japan (1 port, 1 terminal) Kitakyushu	1,100
Saudi Arabia (1 port one terminal) Dammam	1,800
India (5 ports, 5 terminals) Kolkata Mumbai Kakinada Chennai Tuticorin	8,500 4,800 200 1,500 450
Belgium (2 ports, 7 terminals Zeebrugge (2) Antwerpen (5)	1,900 12,100
Italy (2 ports, 3 terminals) Venice Genoa (2)	420 2,350
Portugal (1 port, 1 terminal) Sines	1,700
Turkey (1 port, 1 terminal) Mersin	2,600
Argentina (1 port, 1 terminal) Buenos Aires	1,100
Panama (1 port, 1 terminal) Panama	450
Columbia (1 port, 1 terminal) Buenaventura	1,400
Total: 24 ports and 34 terminals	

48 The World as Military Training Ground

"My next concern was to defend this piece of real estate. We had no army … How were we to build up some defence forces quickly, however rudimentary?"

– Lee Kuan Yew, 2000, p. 6

"'Join the Army and see the world'. Well if you are a regular national serviceman or even a citizen soldier with the SAF, you literally see the world …."

– National Cadet Corps Training Manual Book, 2016, chapter 3

Singapore Armed Forces Training Bases Throughout the World, 2015

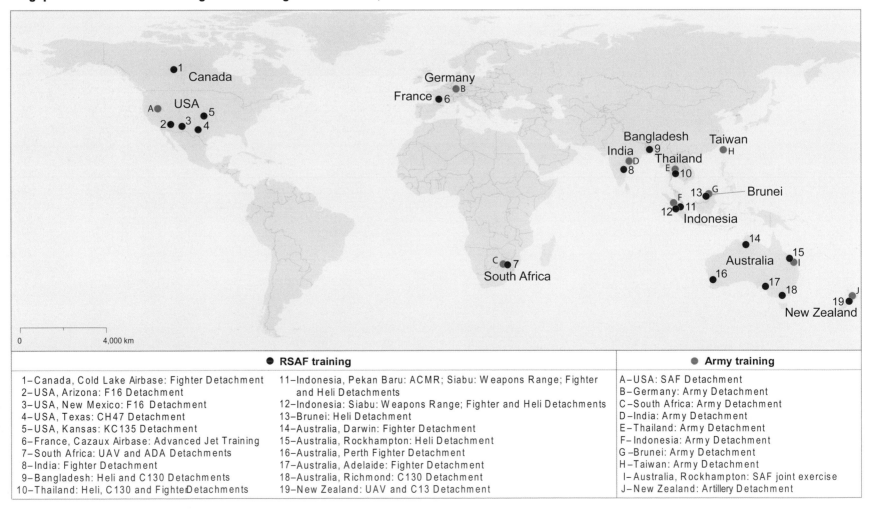

● RSAF training		● Army training
1–Canada, Cold Lake Airbase: Fighter Detachment	11–Indonesia, Pekan Baru: ACMR; Siabu: Weapons Range; Fighter and Heli Detachments	A–USA: SAF Detachment
2–USA, Arizona: F16 Detachment	12–Indonesia: Siabu: Weapons Range; Fighter and Heli Detachments	B–Germany: Army Detachment
3–USA, New Mexico: F16 Detachment	13–Brunei: Heli Detachment	C–South Africa: Army Detachment
4–USA, Texas: CH47 Detachment	14–Australia, Darwin: Fighter Detachment	D–India: Army Detachment
5–USA, Kansas: KC135 Detachment	15–Australia, Rockhampton: Heli Detachment	E–Thailand: Army Detachment
6–France, Cazaux Airbase: Advanced Jet Training	16–Australia, Perth Fighter Detachment	F–Indonesia: Army Detachment
7–South Africa: UAV and ADA Detachments	17–Australia, Adelaide: Fighter Detachment	G–Brunei: Army Detachment
8–India: Fighter Detachment	18–Australia, Richmond: C130 Detachment	H–Taiwan: Army Detachment
9–Bangladesh: Heli and C130 Detachments	19–New Zealand: UAV and C13 Detachment	I–Australia, Rockhampton: SAF joint exercise
10–Thailand: Heli, C130 and Fighter Detachments		J–New Zealand: Artillery Detachment

In the introductory chapter of his *From Third World to First* (2000), Lee Kuan Yew wrote that when Singapore became independent in 1965, his major concerns were that the young republic was first, "to get international recognition" and second, to be able to defend itself; his "third biggest headache" was the economy – "how to make a living for our people?" (p. 7). It did not take more than a decade or two for his first and third "headaches" to be healed. And it would not be an exaggeration to say that his second major concern has equally been resolved, largely through the "globalisation" of the Singapore Armed Forces (SAF).

At first, this meant consulting and collaborating as much as possible with Commonwealth allies, such as the British, Australian and New Zealanders and, with the Israelis. Thanks to the country's rapid economic growth, the government was able to gradually provide all three branches of the SAF (i.e., the army, air force and navy), with the means to purchase the necessary hardware and to offer its military personnel proper training conditions. While, during the early post-independence years, Singapore trainees were dispatched primarily in the region – with the jungles of Brunei representing a key location – their deployment has since become nearly global. The same can be said of the SAF's participation in peace support and disaster response operations.

Singapore still collaborates closely with Malaysia, Australia, New Zealand and the UK, in the context of the 1971 FPDA (Five Power Defence Arrangement), but has established friendly military relations with several other countries. Among those hosting the SAF personnel for their training, are five Asian countries – Brunei, Indonesia, Thailand, Taiwan and India – as well as South Africa and the USA. The deployment for training purposes of the Republic of Singapore Air Force (RSAF) is even more global. Training sites are found in at least 11 countries, including the USA and Canada. In the former, RSAF squadrons are active in four different bases, while in Canada a fighter squadron is training in: Cold Lake, central Alberta. Finally, since 1998, an RSAF fighter squadron has been training in France's Cazaux airbase.

If one were to map Singapore's participation in peace keeping and humanitarian operations the result would be even more impressive. Since the SAF's first overseas deployment for humanitarian assistance was carried out in Bangladesh (then East Pakistan) in 1970, its presence has been significant in numerous locations. Notable deployment of Singaporean servicemen and equipment have occurred in Namibia in 1989, Iraq and Kuwait in 1991, Cambodia in 1993, Timor-Leste on numerous occasions between 1999 and 2003 as well as between 2008 and 2012, Iraq again between 2003 and 2008, Afghanistan since 2007, and in the Gulf of Aden since 2009. Among the more significant humanitarian interventions of the SAF are those that took place in Taiwan in 1999 following an earthquake, the Indonesian province of Aceh in December 2004, after the earthquake and gigantic tsunami, the USA in September 2005, following the onslaught of hurricane Katrina on New Orleans, and where the RSAF sent Chinook helicopters attached to a squadron training in a Texas airbase, and, finally, New Zealand, following the February 2011 earthquake in Christchurch.

The assertion that Singapore is going global can be supported, particularly when it is made by the SAF. Nevertheless, the SAF still maintains a strong foothold within Singapore itself: about 16 per cent of all land is allocated to the military or under restricted access (i.e., reserved for military purposes).

Significant plots of land belong to SAFTI Military Institute in Tengah, Sembawang Air Base, home to the RSAF's rotary wing community, Paya Lebar Air Base as well as Changi Air Base (West). Pulau Tekong is completely designated as a military area, and houses the Basic Military Training Camp.

Military Area in Singapore, 2015

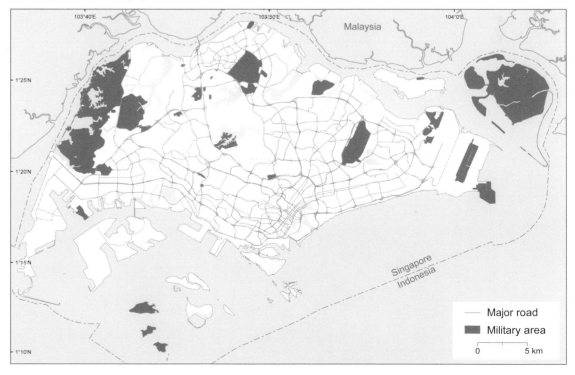

Major road

Military area

0   5 km

Chapter 7

Perpetuating Territorial Transformation and Production

"The Master Plan is the statutory land use plan which guides Singapore's development in the medium term, over the next 10 to 15 years. It is reviewed once every five years, and translates the broad long-term strategies as set out in the Concept Plan into detailed implementable plans for Singapore. It shows the permissible land use and density for every parcel of land in Singapore."

– Urban Redevelopment Authority, Master Plan 2003 website

"It is essential to understand that space precedes territory. … Space represents the "original prison", territory is the prison that men give themselves."

– Translated from Raffestin, 1980, p. 129

"Topophilia takes many forms and varies greatly in emotional range and intensity … the fondness for place because it is familiar, because it is home and incarnates the past, because it evokes pride of ownership or of creation."

– Yi-Fu Tuan, 1974, p. 247

Proposed as early as 1974 by the French philosopher, Henri Lefebvre, the concept of production of space, physical and social, became popular, even fashionable amongst geographers, particularly after his work was translated into English in 1994. Less known but perhaps more fundamental were contributions to the debate on the issue by Raffestin and Bresso (1979) and Raffestin (1980), which made a distinction between space and territory. "Territory", they suggested, is space in which labour has been invested, and territoriality is the relation that people have with space in which they have invested labour as well as emotions. This definition can be related to the one proposed by Yi-Fu Tuan (1974) for topophilia. Redefining and monitoring territoriality and curtailing individual and communal territoriality therefore provide ultimate forms of control.

By constantly "replanning" the rules of access to space, the Singaporean State is thus redefining territoriality, even in its minute details. In this manner it is able to consolidate its control over civil society, peacefully and to an extent rarely known in history. One of the fundamental consequences of this constant and careful planning of land allocation and function is that it enhances the state's unquestionable authority. Notwithstanding occasional dissatisfaction on the part of the citizenry over land use decisions, the fact that these decisions are susceptible of being revised, or adapted, or reviewed, modified and even postponed is both reassuring and worrying. The only solution for the concerned citizen seems to be to trust the system and its planners, who, as Singapore's development achievements have shown for more than a half-century, have generally delivered the goods!

View of the Padang Atrium from the Upper Link Bridge. Courtesy of National Gallery Singapore

49 From Master Plan to Revised Model

There lies one of the more paradoxical elements of citizen participation in Singapore's permanent transformation. Individuals can share or attempt to share in the financial "rewards" of constant redevelopment, but they are not requested nor expected to contribute to the decision making process.

Master Plan

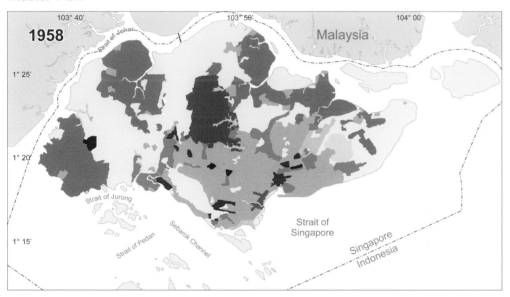

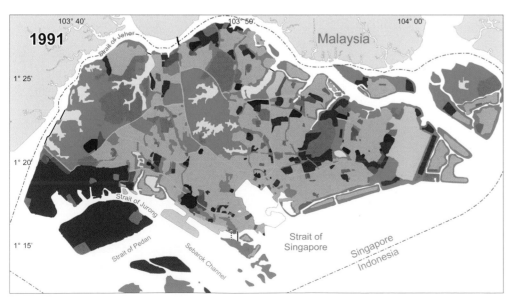

The tendency to first plan territorial functions and then appeal to the population to accept them was already implicit when the First Master Plan was adopted in 1958 after several years of preparation. It was meant to be the basic reference for planning Singapore's development until 1972. But well before that date, the political conditions necessary for its implementation had evolved considerably. The plan was subsequently revised on a five-year basis, and the 1975, 1980 and 1985 versions were substantially different from the original. Another document, the Concept Plan, which was first drafted in 1967 and bore the imprint of the leaders of the newly independent republic, gradually superseded the Master Plan. Adopted in 1970, it defined the real planning goals to be pursued until 1992. Less rigid and more vague than the Master Plan, the Concept Plan still clearly revealed the futuristic views of the Singapore planners and their firm resolution to continue to shape the territory and its occupants to fit their vision of the model state.

The 1991 Master Plan, derived from the Concept Plan as revised in 1991, is substantially different from its 1958 predecessor. It is much more detailed and at the same time more ambitious, particularly with regards to land reclamation,

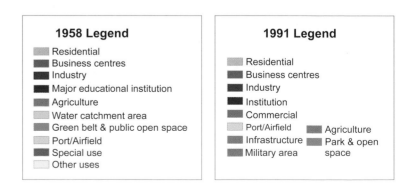

1958 Legend	1991 Legend	
Residential	Residential	
Business centres	Business centres	
Industry	Industry	
Major educational institution	Institution	
Agriculture	Commercial	
Water catchment area	Port/Airfield	Agriculture
Green belt & public open space	Infrastructure	Park & open
Port/Airfield	Military area	space
Special use		
Other uses		

industrial development and population redistribution. Even if the Concept Plan was revised again in 2001 and in 2005, the current Master Plan adopted in 2014 – replacing the ones adopted in 1998 and 2003 – and the futuristic one for 2030 follow the lead of the 1991 version, while being more detailed.

Beyond these Master Plans, which can be consulted on the Urban Redevelopment Authority website, more specific plans, whether thematic or local, are also made available to the public.

In this way Singaporeans can, in principle, learn what is likely to happen to any given piece of land or block of flats on the island and make appropriate personal financial plans. There lies one of the more paradoxical elements of citizen participation in Singapore's permanent transformation. Individuals can share or attempt to share in the financial "rewards" of constant redevelopment, but they nevertheless are neither requested nor expected to contribute to the decision-making process.

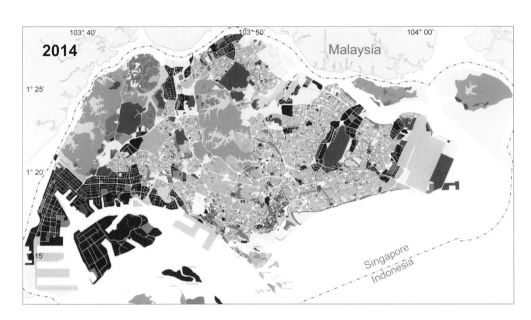

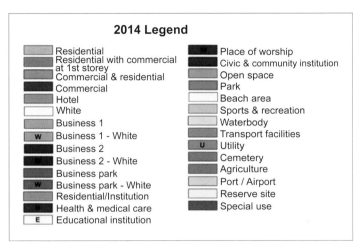

2014 Legend

- Residential
- Residential with commercial at 1st storey
- Commercial & residential
- Commercial
- Hotel
- White
- Business 1
- W Business 1 - White
- Business 2
- W Business 2 - White
- Business park
- W Business park - White
- Residential/Institution
- H Health & medical care
- E Educational institution
- W Place of worship
- Civic & community institution
- Open space
- Park
- Beach area
- Sports & recreation
- Waterbody
- Transport facilities
- U Utility
- Cemetery
- Agriculture
- Port / Airport
- Reserve site
- Special use

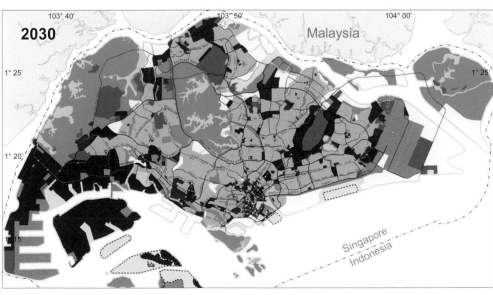

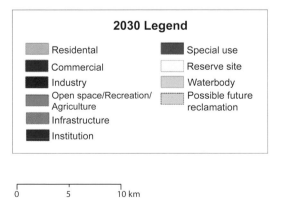

2030 Legend

- Residental
- Commercial
- Industry
- Open space/Recreation/ Agriculture
- Infrastructure
- Institution
- Special use
- Reserve site
- Waterbody
- Possible future reclamation

0   5   10 km

"There is nothing wrong in changing if it is in the right direction", *Winston Churchill*

50 Following the Plan

"We must therefore maximise the value creation from our land."

– Ministry of Trade and Industry, 2003, p. 105

When compared to the 2003 Master Plan map (not illustrated here), the 2005 Land Use map reveals the extraordinary degree to which actual land use conformed to the plan. To a large extent, the same can be said of the 2014 Master Plan map (p. 128) when compared with the 2015 Land Use map presented here, all these being drawn from official sources.

However, the two land use maps are more revealing on some specific issues. Notwithstanding the problem posed by the discrepancies in the land use categories utilised in the 2005 and 2015 maps, the latter also relies on a more detailed pattern of representation, these issues can be listed as follows.

One of the more evident ones concerns the amount of land that is de facto allocated to military use (p. 124). To the substantial areas utilised by the major air bases – starting from the western side of the island, Tengah, Sembawang, Paya Lebar and Changi – and the two naval bases located in Tuas and Changi – must be added several other locations across the island, in particular most of the very large Western Water Catchment district and, of course, all of Pulau Tekong, which by itself covers some 25 km².

At least four additional features are illustrated somewhat more clearly. First, the relative importance of land devoted to green spaces, whether categorised as recreation land, park or wooded area. Second, the manner in which residential areas have nearly become a single contiguous mass centered on the hilly and wooded Bukit Timah Nature Reserve. Third, the stages of land reclamation, distinguishing between recently reclaimed land and what is still to come. Fourth, the still largely extensive category designated as "open space" in 2005 and, in 2015, "grass, turf, shrubs and general vegetation". Both of these actually refer to land still to be developed.

All land use planning in Singapore follows the planners' dominant goal to maximise value creation from the land. This is achieved increasingly in a context of intense competition with Malaysia. Singapore has first expanded

Land Use Map

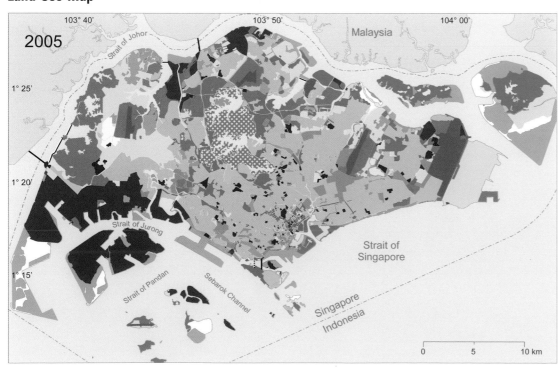

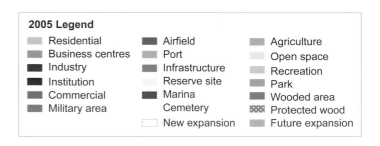

2005 Legend

Residential	Airfield	Agriculture
Business centres	Port	Open space
Industry	Infrastructure	Recreation
Institution	Reserve site	Park
Commercial	Marina	Wooded area
Military area	Cemetery	Protected wood
	New expansion	Future expansion

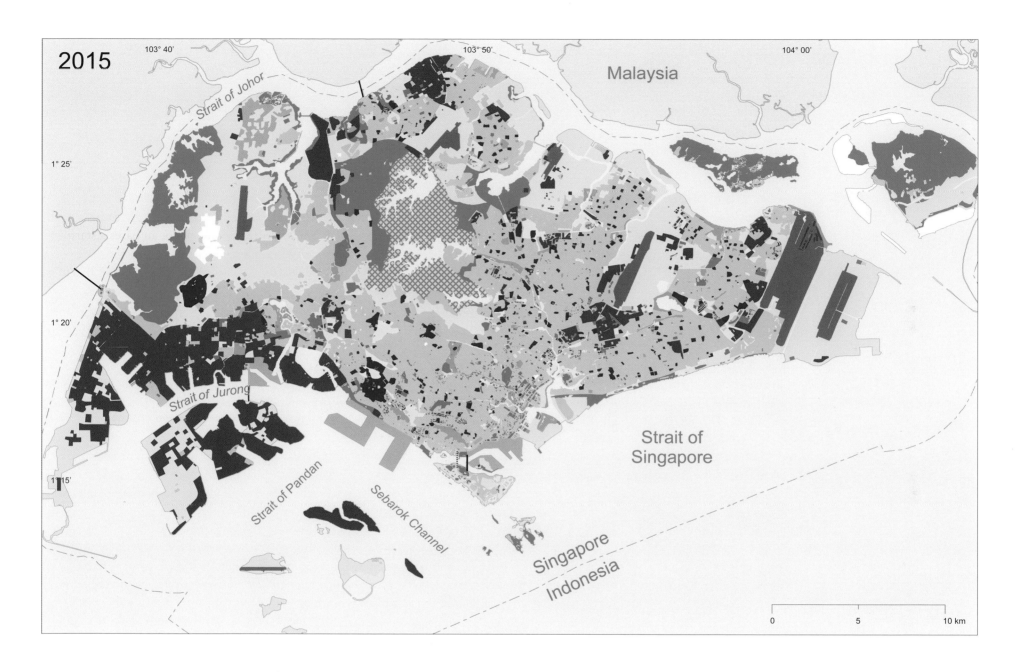

2015 Legend

Built-up area	New expansion
Industry	Agriculture
Institution	Grass / Turf / Shrubs & General vegetation
Public / Commercial	Recreation
Airfield	Park / Garden
Port	Wooded / Forested area
Cemetery	Protected wood
	Waterbody

its own territory which in some locations, such as in its extreme southeast and southwest portions, has brought it closer to Malaysian territory (p. 20). It has even relocated a fair portion of its industrial infrastructure in the state of Johor and is now faced with Johor's own industrial boom and territorial expansion, just across the Johor Strait (p. 118). Following the plan will likely test Singapore's innovation capacities to its limit.

Conclusion: The Moveable Stage

Doesn't the permanent transformation of Singapore's territory by the state represent an unprecedented effort at creating and maintaining legible landscapes?

There is much more to the transformations of the territory of Singapore than can be illustrated here.

First, the discussion has only dealt with a relatively small number of topics, with many other potentially revealing ones being left aside. These include, for example, the proliferation of automated teller machines (ATMs), which the Singaporean banks have been disseminating throughout the island by the thousands, including in its few remaining remote corners. Such a map would illustrate, even more than that of shopping malls (p. 88), the extent to which Singapore has become a hyper-consumer society, one where "life is not complete without shopping", as former Prime Minister Goh Chok Tong once said (Chua, 2003).

Second, changing distributions have mostly been illustrated on a country-wide scale. It is quite obvious, as suggested in the introduction, that the same changes are occurring on a more local scale: at the level of the district, or the neighbourhood, such as Chinatown, or the street, such as Bugis street, etc. Other examples are provided by bus routes, such as that of bus no 7, which has been rather constant since it was opened in 1979, and to which no significant changes have been made since 1991. At this level photographs taken at different points in time would illustrate the magnitude of the territorial transformations.

Third, data remain sparse in some areas. This is the case with a particularly sensitive issue that is part and parcel of the Singaporean permanent transformation: the function and distribution of foreign workers in the country. Although we have been able to map the location of the majority of the dormitories in which these workers are increasingly encouraged to move (p. 44), data were surprisingly hard to come by.

Nor does the book delve into the detailed political economy of planned environmental transformation. This is not because the topic is unimportant but rather that in social science investigations, as in all forms of scientific inquiry, it is often indispensable to isolate the object of study in order to make it more readily understandable and not to dilute and weaken the approach. The changes undergone by the territory of the island republic are only one part of an extremely ambitious project: the Singapore experiment, the Singapore revolution, as suggested in the introduction, has territorial, financial, political, and even ideological ramifications that extend far beyond the borders of the 720 km^2 city-state. These are definitely beyond the scope of the present study.

Our original hypothesis was that, among members of a community, even a national community, territorial instability – or better still territorial alienation – can become tools of social and political control and alienation. There lies one of the more paradoxical elements of citizen participation in Singapore's permanent territorial revolution, a revolution entirely monitored from above. Individuals can share or attempt to share in the financial "rewards" of constant redevelopment, but they nevertheless are neither invited nor expected to contribute to the decision-making process, since their individual and collective territoriality does not seem to weigh much in the balance.

This leads to the following questions and hypothetical interpretations. Isn't the relentless overhaul of Singaporean living space – nearly always considered as a *fait accompli*, yet always subject to being revised by the state – leading to territorial alienation among the City-State's citizens and permanent residents? Could this alienation possibly contribute to political resignation among the City-State's citizens and permanent residents? Positive answers to these two questions lead to the following dual proposal. First, permanent and unquestionable remoulding of individual and communal territorial markers is a tool of political control, whether used consciously or not. Second, since constant remoulding and remodeling leaves little space for the urban imaginative field in Singapore, the soft city – elements of the city which can be the object of topophilia, that to which citizens can emotionally relate and contribute to build – has a long way to go before it can challenge the hard one.

As stated in the introduction, the validity of these proposals cannot be readily measured. Nevertheless, we feel we have been able to show that the territorial hypothesis has an explanatory potential worthy of further investigation and debate.

In fact, in view of the pace at which the Singapore environment, the stage of a unique experiment in political and social construction, continues to change, the country remains an appropriate testing ground of an even more explicit question. This question could have been raised by James Scott in his 1998 seminal book *Seeing as a State*: Doesn't the permanent transformation of Singapore's territory by the state represent an unprecedented effort at creating and maintaining legible landscapes?

Appendix 1: Dates and Events

2nd C	In his *Geography*, Ptolemy of Alexandria mentions the port of Sabara (or Sabana), which could have been located on Singapore island.
14th C	Singapore Island is the site of an influential Malay maritime city.
1819	Stamford Raffles establishes a trading post in the name of the East India Company.
1824	The British formally acquire the island from the Sultan of Johor.
1826	The Straits Settlements, which includes Singapore, is formed.
1865	Transfer of control over the Straits Settlements from the East India Company to the British Colonial Office.
1869	Opening of the Suez Canal.
1887	Straits Trading Company Limited is established
1899	Construction of Singapore-Johor Straits Railway is approved
1913	Singapore Harbour Board is formed
1923	Beginning of the development of the Sembawang air and naval base.
1929	Opening of the Seletar air base.
1938	Opening of the King George VI Dry Dock.
1938	Construction of the Tengah air base.
1942	Singapore is captured by the Japanese Armed Forces (15 February).
1945	Departure of Japanese occupants (September).
1958	Singapore's first Master Plan is adopted.
1959	Internal autonomy granted by the British; the head of the People's Action Party (PAP) Lee Kuan Yew is the first Prime Minister.
1963	Integration of Singapore into the newly created Federation of Malaysia.
1965	Singapore leaves the Federation and becomes an independent republic; Lee Kuan Yew remains Prime Minister.
1967	Creation of the Association of Southeast Asian Nations (ASEAN), which includes Singapore.

1970	Adoption of the first Singapore Concept Plan, called the Ring Concept Plan.
1971	Definite closure of all British military installations.
1981	Opening of Changi airport.
1990	Goh Chok Tong succeeds Lee Kuan Yew as Prime Minister. The latter remains with the government as Senior Minister.
1997	Asian Financial Crisis.
2003	Adoption of a new Master Plan, after the 1958, 1975, 1980, 1985, 1991 and 1998 ones.
2003	SARS outbreak.
2004	Lee Hsien Loong (son of Lee Kuan Yew), succeeds Goh Chok Tong as head of the PAP and as Prime Minister. Goh Chok Tong remains with the government as Senior Minister, while Lee Kuan Yew becomes Minister Mentor.
2005	Adoption of latest revised Concept Plan, after the 1971, 1991 and 2001 revisions.
2006	The Global Competitiveness Report ranks Singapore no. 5, ahead of the US, and only surpassed by four northern European countries.
2008	Adoption of a new Master Plan.
2010	Hosting Singapore's first casino, Marina Bay Sands hotel opens.
2010	Singapore has become a major medical tourism hub, with well over 200,000 foreigners coming to the City-State to seek medical services.
2012	The country facilitates the recruitment of foreign students in a move to improve its position in the global knowledge world.
2013	Riots occur in Little India, the first in Singapore since 1969.
2014	Adoption of a new Master plan.
2015	Singapore tops the OECD's global school performance rankings, based on 15-year-old students' average scores in mathematics and science across 76 countries. It is also ranked by the World Bank as the world's easiest place to do business.
2015	On 23 March, Lee Kuan Yew, prime minister from 1959 to 1990, dies.
2015	On 9 August, Singapore celebrates 50 years of independence as a full-fledged republic.

Appendix 2: Singapore 1959-2015

	1959	2015
Surface area	581.5 km²	719.0 km²
Population (Residents only)[a]		3,902,700
Population (Total)[a]	1,587,200	5,535,000
Major ethnic groups (in 1957 and 2000) (%)		
Chinese	75.4	76.2
Malays	13.6	15.0
Indians	9.0	7.4
Others	2.0	1.4
Birth rate per thousand	32.9	9.8
Mortality rate per thousand	6.4	4.7
Infant mortality rate per thousand	36.0	1.8
Life expectancy	64[b]	83
Annual population growth (%)	4.0	1.2
Unemployment (%)	13.2	3.6
Literacy rate (15 years and above) (%)	52.2[b]	97.0
GDP (millions SGD)	2,150[c]	390,100[e]
External trade (millions SGD)	6,811[d]	977,000[e]
SGD to the USD	~ 0.33	~0.66

a. Total population comprises Singapore residents and non-residents. Resident population comprises Singapore citizens and permanent residents.
b. 1957
c. 1960
d. 1965
e. 2014

Sources : *United Nations Demographic Yearbooks* and http://www.singstat.gov.sg/

Bibliography

I. Sources for the Base Maps and the Tables

(Note: A number of the websites listed here were consulted for the preparation of the
2008 edition of the atlas. Many of them are not active anymore.)

Department of Statistics. *Yearbooks.*
Environmental Systems Research Institute (ESRI). *Digital Chart of the World.*
Google Earth, Singapore, several dates since 2006. Image ©2016 Digital Globe.
Google Maps, Singapore, several topics in 2016. Map Data ©2016 Google.
Ministry of Defence, Mapping Unit (1958, 1966). Topographic 1: 63,360.
_____ (1974, 1987). Topographic 1: 50,000.
_____ (1988). *Road Map.* 1: 25,000 (4 sheets).
_____ (2011). Topographic 1: 50,000.
OpenStreetMap (OSM), 2015.
Singapore Guide and Street Directory (1958).
Singapore Street Atlas, several years.
Singapore Street Directory, several years since 1988.

1. Singapore: A Global Strategic Location
ESRI. *Digital Chart of the World.*
Map adapted from: De Koninck, Rodolphe (2006). *Singapour: La cité-État ambitieuse.*
    Paris: Belin. Map 1, p. 19.
Rodrigue, Jean-Paul, et al. (2013). *The Geography of Transport Systems.* London and
    New York: Routledge.
The Geography of Transport Systems (2013). "Detailed Maps." http:/www.
    people.hofstra.edu/geotrans/eng/media_maps.html.
Turnbull, C.M. (2009). *A History of Singapore, 1819–2005.* Singapore: NUS Press.

2. Singapore: A Regional Strategic Location
Map courtesy of Quiex, on the Wikimedia Commons.
Miksic, John (2013). *Singapore and the Silk Road of the Sea, 1300–1800.* Singapore:
    NUS Press.
Sandhu, K.S. and Wheatley, Paul, eds. (1989). *Management of Success: The Moulding of
    Modern Singapore.* Singapore: Institute of Southeast Asian Studies.

3. Singapore in the Midst of Historical Trade Centres
ESRI. *Digital Chart of the World.*
Map adapted from De Koninck (2006), op. cit., Map 2, p. 23.
Miksic, John N. (2004). "14th Century Singapore: A Port of Trade." In John N. Miksic
    and Cheryl-Ann Low (eds.), *Early Singapore 1300s–1819: Evidence in Maps, Text
    and Artefacts.* Singapore: Singapore History Museum, pp. 41–54.

4. Singapore as a Migrants' Haven
Department of Statistics Singapore. http://www.tablebuilder.singstat.gov.sg/
    publicfacing/mainMenu.action [accessed March 2016]
Index Mundi (2015). "Singapore Net Migration Rate." Last modified June 30, 2015.
    http://www.indexmundi.com/singapore/net_migration_rate.html [accessed July
    2015].
Martin, Robert Montgomery (1839). *Statistics of the Colonies of the British Empire.*
    London: W.H. Allen.
Tan, Jeanette (2014). "More foreign workers would recommend Singapore as a good
    place for work: MOM survey." Yahoo! News. Last modified July 7, 2014. https://
    sg.news.yahoo.com/more-foreign-workers-would-recommend-singapore-as-a-
    good-place-for-work--mom-survey-114310058.html [accessed July 2015].
Yeoh, Brenda and Lin Weiqiang (2012). "Rapid Growth in Singapore's Immigrant
    Population Brings Policy Challenges." Migration Policy Institute. Last modified
    April 3, 2012. http://www.migrationpolicy.org/article/rapid-growth-singapores-
    immigrant-population-brings-policy-challenges.

5. Contemporary Singapore.
ArcGIS Online World Topographic Map.
Derived data from the OpenStreetMap (OSM) dataset.
ESRI (2010). "ArcGIS Online World Topographic Map." Last
    modified May 27, 2010. https://www.arcgis.com/home/item.
    html?id=a72b0766aea04b48bf7a0e8c27ccc007.
Google Earth, Image ©2016 Digital Globe.
Housing and Development Board (HDB) (2010). "Location of HDB Development." In
    Key Statistics for 2009/2010 (a HDB Annual Report for financial year 2009/2010),
    p. 9. http://www88.hdb.gov.sg/eBook/AnnualReport/Statistics2010.html
Leow Bee Geok (2001). *Census of Population 2000 Statistical Release 4: Geographic
    Distribution and Travel.* Singapore: Department of Statistics.
National Aeronautics and Space Administration (NASA) (2016). "U.S. Releases
    Enhanced Shuttle Land Elevation Data." Last modified May 16, 2016. http://www2.
    jpl.nasa.gov/srtm/index.html.
Urban Redevelopment Authority (2015). "Master Plan 2014." http://www.onemap.sg/
    index.html?PopulationQuery
https://www.openstreetmap.org/relation/536780

7. Stretching the Land
Google Earth. "Singapore." 1°21›05.52»N and 103°49›54.1»E. September 18, 2015
    and before.

Koh, Tomy and Lin, Jolene (2006). "The Land Reclamation Case: Thoughts and Reflections." Singapore Year Book of International Law and Contributors 10, (2006): 1–7. https://lkyspp.nus.edu.sg/wp-content/uploads/2013/04/pa_tk_The-Land-Reclamation-Case-Thoughts-and-Reflections-_2007.pdf

Lee Chiang Fong. "How much reclaimed land has been added to Singapore, excluding Jurong Island and other offshore islands, since 2000?" *Quora*. Last modified December 19, 2014. https://www.quora.com/Singapore/How-much-reclaimed-land-has-been-added-to-Singapore-excluding-Jurong-Island-and-other-offshore-islands-since-2000

Leow (2001), op. cit.

Map adapted from: De Koninck (2006), op. cit., Map 7, p. 72.

Ministry of Defence, Mapping Unit (1958, 1966). Topographic 1: 63,360.

_____(1974, 1987). Topographic 1: 50,000.

_____(1958). Topographic 1: 25,000 (7 sheets).

_____ (2011). Topographic 1: 50,000.

One Map Singapore. Singapore Historical Map. 1957, 1981, 1991 and 2000 versions. http://hm.onemap.sg/

Tan, Theresa (2015). "S'pore's largest reclamation project begins." *Straits Times*. Last modified September 27, 2015. http://www.straitstimes.com/singapore/spores-largest-reclamation-project-begins

Wong Tai-Chee and Yap Lin-Ho, Adriel (2004). *Four Decades of Transformation: Land Use in Singapore, 1960–2000*. Singapore: Eastern University Press, p. 121.

7.   Searching for Land

Comaroff, Joshua. "Built on Sand: Singapore and the New State of Risk." Harvard Design Magazine. http://www.harvarddesignmagazine.org/issues/39/built-on-sand-singapore-and-the-new-state-of-risk [accessed August 2015].

Koh, Tomy and Lin, Jolene (2006). "The Land Reclamation Case: Thoughts and Reflections." *Singapore Year Book of International Law and Contributors* 10 (2006): 01-07. https://lkyspp.nus.edu.sg/wp-content/uploads/2013/04/pa_tk_The-Land-Reclamation-Case-Thoughts-and-Reflections-_2007.pdf

Loh, Andrew (2014). "S'pore Thirst for Sand Again in the News." The *Online Citizen*. Last modified April 5, 2014. http://www.theonlinecitizen.com/2014/04/spores-thirst-for-sand-again-in-the-news/ [accessed August 2015].

Milton, Chris (2010). "The Sand Smugglers." Foreign Policy. Last modified August 4, 2010. http://foreignpolicy.com/2010/08/04/the-sand-smugglers/ [accessed August 2015].

The Economist (2015). "Such Quantities of Sand." Last modified February 28, 2015. http://www.economist.com/news/asia/21645221-asias-mania-reclaiming-land-sea-spawns-mounting-problems-such-quantities-sand [accessed August 2015].

United Nations. *UN Commodity Trade Statistics, 2014*: UN Comtrade Database. http://comtrade.un.org/

Yoong, Sean et al. (2001). "Where does Singapore come from?" Through the Sandglass. Last modified September 1, 2011. http://throughthesandglass.typepad.com/through_the_sandglass/2011/09/where-does-singapore-come-from.html [accessed August 2015].

http://na.unep.net/geas/getUNEPPageWithArticleIDScript.php?article_id=110

http://www.huffingtonpost.com/2011/08/22/sand-dredging-singapore-construction-environment-cambodia_n_932840.html [accessed August 2015].

8.   Pulau Tekong as New Frontier

Borschberg, Peter, ed. (2014). *The Memoirs and Memorials of Jacques de Coutre: Security, Trade and Society in 16th- and 17th-century Southeast Asia*. Singapore: NUS Press, trans. Roopanjali Roy.

Borschberg, Peter, ed., annotated and introduced (2015). *Jacques de Coutre's Singapore and Johor 1594–1625*. Singapore: NUS Press.

Chen Poh Seng and Lee Leng Sze (2012). *A Retrospect of the Dust-Laden History. The Past and Present of Tekong Island in Singapore*. Singapore: World Scientific Publishing.

Google Earth, several dates since 2006.

Ministry of Defence, Mapping Unit (2011). Topographic 1: 50,000.

OpenStreetMap (2015).

Singapore Street Directory (2015).

Survey Department, Federation of Malaya, 1958. Topographic 1:25,000 (sheet 134d).

"Tekong Revisited." *Straits Times*, 24 May 2009.

http://eresources.nlb.gov.sg/infopedia/articles/SIP_1009_2010-05-14.html

9.   Collecting and Stocking Water

De Koninck, Rodolphe (1992). *Singapour. Un Atlas de la Révolution du Territoire/ Singapore. An Atlas of the Revolution of Territory*. Montpellier: RECLUS, p. 41.

Google Earth, several dates since 2006.

Ministry of Defence, Mapping Unit (1958). Topographic 1: 63,360.

_____ (1987). Topographic 1: 50,000.

_____ (2011). Topographic 1: 50,000.

Wong and Yap (2004), op. cit., Figure 5.1B.

http://www.pub.gov.sg/water/localcatchment/Pages/keepclean.aspx

10.   Diversifying Water Supply Sources

Google Earth, several dates since 2006.

Public Utilities Board (PUB), Singapore's National Agency.

Photograph by Edwin Lee (Flickr user edwin11) shared on a CC-BY 2.0 license.

http://www.pub.gov.sg/water/Pages/singaporewaterstory.aspx

http://www.pub.gov.sg/mpublications/Documents/Fact%20sheet%20on%20Deep%20Tunnel%20Sewerage%20System%20With%20IWMF.pdf

http://s211lec01earthisours.pbworks.com/w/page/49238068/Norisah%20Zul%20
  Hijah%20(1)

https://blogs.ntu.edu.sg/hp331-2014-51/?page_id=25

11. The Garden City

De Koninck (1992), op. cit., pp. 49 and 105

_____ (2006), op. cit., p. 98.

Bird, Michael et al. (2004). "Evolution of the Sungei Buloh-Kranji mangrove coast,
  Singapore." *Applied Geography* 24: 181–98.

Google Earth, several dates since 2006.

Hill R.D. (1977). "The Vegetation Map of Singapore." *Journal of Tropical Geography* 45:
  27–33.

Ministry of Defence, Mapping Unit (1958). Topographic 1: 63,360.

_____ (1958). Topographic 1: 25,000 (7 sheets).

_____ (1987). Topographic 1: 50,000.

_____ (1988). *Road Map.* 1: 25,000 (4 sheets).

_____ (2011). Topographic 1: 50,000.

National Parks Board (2006). http://www.nparks.gov.sg/publishing.asp

National Parks © 2016 National Parks Board. All rights reserved. https://www.nparks.
  gov.sg/gardens-parks-and-nature

Singapore Guide and Street Directory (1958).

Singapore Street Directory (1988).

Singapore Street Directory (2007).

Singapore Street Atlas (2005–06).

The data.gov.sg Portal © 2016 Government of Singapore. https://data.gov.sg/dataset/
  parks

Urban Redevelopment Authority (URA) ©2016 OneMap. "Parks and Waterbodies
  Plan." http://www.onemap.sg/index.html

12. The Sea in the City

Coleman, G.D. (1837). "Map of the Town and Environs of Singapore." In Newbold
  (1839).

Map adapted from: De Koninck (2006), op. cit., Map 3, p. 27.

Newbold, T.J. (1839). *Political and statistical account of the British settlements in the
  Straits of Malacca*. London: John Murray.

"Singapore." 1°17›12.54»N and 103°52›1.47»E. Google Earth. March 26, 2001 and
  September 29, 2014.

Singapore Street Atlas (2005–06). Singapore: Periplus Editions.

13. Spreading Out the Population

Census of Population 1980 Singapore, Release No.4, Economic Characteristics, p. 1,
  Map: http://hm.onemap.sg/1981.

Census of Population 2000, Administrative Report, p. 40,
  Map: Map: http://hm.onemap.sg/2000.

Census of Population 2010: Advance Census Release, Table A1: Resident Population
  by Age Group, Ethnic Group, Sex and Residential Status.

Department of Statistics. *Yearbooks.*

Map adapted from: De Koninck (2006), op. cit., Maps 9, 10 and 12, pp. 80, 81 and 83.

Map: Google Earth 7.1.5.1557 (April 20, 2015). Singapore. 1° 16' 51.08"N, 103° 49'
  19.8"E, Eye alt 14.09 km. DigitalGlobe 2016.

Neville W. (1965). "The Areal Distribution of Population in Singapore." *Journal of
  Tropical Geography* 20: 16–25.

Report on the Census of Population 1957, p. 17, Map: http://hm.onemap.sg/1957.

Statistics Singapore - Census of Population 2010, Publications.

http://www.singstat.gov.sg/publications/publications-and- papers/cop2010/
  census10_stat_release1

http://www.singstat.gov.sg/publications/publications-and-papers/reference/
  yearbook-of-statistics-singapore

http://www.singstat.gov.sg/docs/default-source/default-document-library/
  publications/publications_and_papers/cop2010/census_2010_release3/
  map1.pdf

14. The Housing Question

De Koninck (1992), op. cit., pp. 90-1.

Housing and Development Board (HDB). *Key Statistics for 2010/2011. Location of
  HDB Developments*. HDB Annual Report 2010/11, p. 9.

Housing and Development Board (HDB) Map Services.

http://www10.hdb.gov.sg/ebook/ar2011/keystats.html

http://www.straitstimes.com/singapore/housing/12000-new-flats-to-be-launched-in-
  november

http://services2.hdb.gov.sg/web/fi10/emap.html

15. Housing Foreign Workers

Ministry of Manpower (2015*). Labour Force in Singapore, 2014*.

Yeoh, Brenda and Lin Weiqiang (2012), op. cit.

http://foreignworkerdormitory.com/foreign-workers-may-be-arranged-at-close-
  offshore- islands-of-singapore [accessed August 2015].

http://foreignworkerdormitory.com/approved-dormitory-list-by-mom-in-
  november-2014 [accessed August 2015].

16. Laying out a New Town: Punggol

Basemap © OpenStreetMap (OSM) contributors (2015).

"Punggol." 1°24›10.48»N and 103°54›30.7»E. Google Earth. July 6, 2004 and July
  23, 2015.

Singapore Historical Map. Map 1991 and Map 1993.

Singapore Street Atlas. Singapore: Periplus Editions, several years.

Singapore Street Directory, several years.

Singapore Land Authority (SLA) Cadastral Land Lot (22-JUL-2014), data.gov.sg.

http://hm.onemap.sg/

https://ref.data.gov.sg/Metadata/OneMapMetadata.aspx?id=Cadastral_Land_
     Lot&mid=236326&t=SPATIAL

https://www.openstreetmap.org/#map=15/1.4071/103.9205&layers=T

http://www.kevinwylee.com/homepage

http://www.punggol.com/

https://en.wikipedia.org/wiki/Punggol_New_Town

17. Private Quarters

De Koninck (1992), op. cit., p. 95.

Housing and Development Board (HDB) Map Services.

Singapore Guide and Street Directory (1958).

Singapore Street Atlas. Singapore: Periplus Editions, several years.

Singapore Street Directory, several years.

http://services2.hdb.gov.sg/web/fi10/emap.html

18. Readjusting the Distribution of Ethnic Communities

Census of Population 2000.

Census of Population 2010.

Leow (2001), op. cit., pp. 43-50.

Khoo Chian Kim (1983). *Census of Population 1980 Singapore.* Singapore: Department
     of Statistics Singapore, pp. 234–44.

19. Rationing Agriculture

Agri-Food and Veterinary Authority of Singapore (2006). "Agrotechnology Parks in
     Singapore."

De Koninck, R. (1975). *Farmers of a City-State. The Chinese Smallholders of Singapore.*
     Montreal: Canadian Sociology and Anthropology Association.

De Koninck, R. (1992), op. cit., p. 53.

Ministry of Defence, Mapping Unit (1958). Topographic 1: 63,360.

_____ (1987). Topographic 1: 50,000.

_____ (2011). Topographic 1: 50,000.

Wong (1989), op. cit.

http://data.gov.sg/ [accessed August 2015].

http://www.ava.gov.sg/explore-by-sections/farms/land-farms/farming-in-singapore
     [accessed April 2016].

http://www.ava.gov.sg/AgricultureFisheriesSector/FarmingInSingapore/
     AgroTechParks/AgrotechnologyParksMap.htm [accessed April 2016].

20. Expanding and Consolidating Industry

De Koninck (1992), op. cit., p. 55.

Google Earth, several dates since 2006.

Jurong Town Corporation. *Annual Reports.*

Ministry of Defence, Mapping Unit (1958). Topographic 1: 63,360.

_____ (1987). Topographic 1: 50,000.

OpenStreetMap (2015).

Singapore Street Directory (2015).

Urban Redevelopment Authority (2003), http://spring.ura.gov.sg/dcd/eservices/
     sop/main.cfm?viewmpview

Wong A.K. and Ooi G.L. (1989). "Spatial Reorganization." In Sandhu and Wheatley
     (eds.), pp. 788-812.

https://www.edb.gov.sg/content/edb/en/industries/industries/electronics.html

21. Jurong: From Mangrove to Industrial Estate to Urban Status

Photograph by Allie Caulfield (Flickr user wm_archiv) shared on a CC-BY 2.0 license.

Ministry of Defence, Mapping Unit (1958). Topographic 1: 25,000 (7 sheets).

_____ (2011). Topographic 1: 50,000.

Singapore Street Atlas. Singapore: Periplus Editions, several years.

Singapore Street Directory (2015).

Urban Redevelopment Authority (2003).

http://spring.ura.gov.sg/dcd/eservices/sop/main.cfm?viewmpview

22. Powering Singapore

Chang et al. (2001). RAI International Exhibition and Congress Centre. Amsterdam,
     The Netherlands, 18-21 June 2001. CIRED Conference Papers.

http://www.pub.gov.sg

http://www.energyportal.sg/Statistics/Energy-in-Number.html

https://ref.data.gov.sg [accessed January 2016]

https://www.ema.gov.sg/ [accessed January 2016]

http://globalenergyobservatory.org/index.php [accessed January 2016]

23. Petroleum Islands

Atlas for Singapore (1988). Singapore: Collins and Longman.

De Koninck (1992), op. cit., p. 63.

Google Earth, several dates since 2006.

Ministry of Defence, Mapping Unit (1958). Topographic 1: 63,360.

_____ (1987). Topographic 1: 50,000.

_____ (2011). Topographic 1: 50,000.

Singapore Street Atlas 2005-2006. Singapore: Periplus Editions.

Singapore Street Directory (2007). Singapore: Periplus Editions.

Singapore Street Directory (2015). Singapore: Periplus Editions.

http://globalenergyobservatory.org/index.php [accessed January 2016]

24. Making Way for Cars
Atlas for Singapore (1979). Singapore: Collins and Longman.
De Koninck (1992), p. 69.
Mahbubani, K. (2014). "The New Singapore Dream." *Commentary* 23: 65–71.
Ministry of Defence, Mapping Unit (2011). Topographic 1: 50,000.
Nelles Verlag (1990). *Singapore* 1: 125,000.
Singapore Street Atlas 2005-2006. Singapore: Periplus Editions.
Singapore Street Directory (1991).
Singapore Street Directory (2007). Singapore: Mighty Minds.
Transport system map © OpenStreetMap (OSM) contributors (2015).
Wildermuth, B. (2015). "Singapore has become a City for Cars, not People." *Commentary* 24: 15–21.
Wong and Yap (2004), op. cit., Figure 2.2.
https://www.openstreetmap.org/#map=12/1.3558/103.8160&layers=T

25. Transporting Workers
Atlas for Singapore (1979). Singapore: Collins and Longman.
De Koninck (1992), op. cit., p. 69.
Land Transport Authority. 2006, in *Projects*. http://www.lta.gov.sg/
Nelles Verlag (1990). *Singapore*. 1: 125,000.
Singapore Street Atlas 2005-2006. Singapore: Periplus Editions.
Singapore Street Atlas 2015. Singapore: Periplus Editions.
Transport system map © OpenStreetMap (OSM) contributors (2015).
https://www.openstreetmap.org/#map=12/1.3593/103.8184&layers=T

26. Places to Pray
De Koninck (1992), op. cit., p. 99.
_____ (2015). "Singapore: Condemned to Change?", *NUSS Commentary* 24: 47–53.
Ministry of Defence, Mapping Unit (1958). Topographic 1: 25 000 (7 sheets).
National University of Singapore (NUS) Libraries, Singapore Places of Worship, April 2016.
Singapore Guide and Street Directory (1958).
Singapore Street Atlas 2005-2006. Singapore: Periplus Editions.
Singapore Street Atlas 2015. Singapore: Periplus Editions.
Singapore Street Directory (1988).
"Singapore (all places of worship)." ©2016 Google Maps.
http://libportal.nus.edu.sg/frontend/ms/sg-places-worship/about-sg-places-worship
https://www.google.ca/maps/place/Singapour/@1.3147308,103.8470191,11z/

27. Places for Burial
After Death Facilities © 2016 Government of Singapore. Data.gov.sg.
De Koninck (1992), op. cit., p. 101.

Harfield, Alan (1988). *Early Cemeteries in Singapore*. London: British Association for Cemeteries in South Asia.
Ministry of Defence, Mapping Unit (1958). Topographic 1: 25 000 (7 sheets).
Singapore Guide and Street Directory (1958).
Singapore Street Atlas 2005-2006. Singapore: Periplus Editions.
Singapore Street Atlas 2015. Singapore: Periplus Editions.
Singapore Street Directory (1988).
"Singapore (grave yard, cemetery, columbarium)." Map Data ©2016 Google.
http://www.roughguides.com/destinations/asia/singapore/northern-singapore/bukit-brown-cemetery/#ixzz43zXN74SI
https://geo.data.gov.sg/afterdeathfacilities/2015/02/02/shp/afterdeathfacilities.zip
https://www.google.ca/maps/place/Singapour

28. Places to Study
De Koninck (1992), op. cit., p. 103.
List of schools in Singapore at Wikipedia
Ministry of Defence, Mapping Unit (1958). Topographic 1: 25 000 (7 sheets).
Ministry of Education (1990). *Directory of Schools and Financial Institutions.*
_____ (2007). In *Education System, Post-secondary*. http://www.moe.gov.sg/corporate/post_secondary.htm
Morais, J.V. and Pothen P.P. (1965). *Educational Directory of Malaysia and Singapore.* Singapore.
Singapore Guide and Street Directory (1958).
Singapore Street Atlas 2005-2006. Singapore: Periplus Editions.
Singapore Street Atlas 2015. Singapore: Periplus Editions.
Singapore Street Directory (1966).
"Singapore (primary schools, secondary schools, higher education schools)." Map Data ©2016 Google.
https://en.wikipedia.org/wiki/List_of_schools_in_Singapore
https://www.google.ca/maps/place/Singapour

29. Places for Recreation
De Koninck (1992), op. cit., p. 105.
Ministry of Defence, Mapping Unit (1958). Topographic 1: 25,000 (7 sheets).
_____ (1988). *Road Map.* 1: 25,000 (4 sheets).
Singapore Guide and Street Directory (1958).
Singapore Street Atlas 2005-2006. Singapore: Periplus Editions.
Singapore Street Atlas 2015. Singapore: Periplus Editions.
Singapore Street Directory (1988).
Singapore Street Directory (2015).

30. Rallying Points: Community Centres
De Koninck (1992), op. cit., pp. 10–9.

Maung T.T. (1957). *The Influence of the Community Centre on its Neighbourhood*. Singapore: University of Malaya (Social Studies).

People's Association. *Annual Reports*.

Singapore Street Atlas 2005-2006. Singapore: Periplus Editions.

Singapore Street Atlas 2015. Singapore: Periplus Editions

http://data.gov.sg/Metadata/OneMapMetadata. aspx?id=COMMUNITYCLUBS&mid=208399&t=SPATIAL [access November 2015].

31.   Rallying Points: Shopping Malls

Chua Beng Huat (2003). *Life is not Complete without Shopping: Consumption Culture in Singapore*. Singapore: Singapore University Press.

Google Earth, several dates since 2006.

Ooi G.L. (1991). "Urban Policy and Retailing Trends in Singapore." *Urban Studies* 4: 585-96.

Singapore Business, 4 February 2016

Singapore Street Atlas 2005-2006. Singapore: Periplus Editions.

Singapore Street Atlas 2015. Singapore: Periplus Editions.

Wildermuth, B. (2015), op. cit.

https://en.wikipedia.org/wiki/VivoCity

https://en.wikipedia.org/wiki/HarbourFront_Centre

https://en.wikipedia.org/wiki/Jurong_Point_Shopping_Mall

https://en.wikipedia.org/wiki/List_of_shopping_malls_in_Singapore

http://sbr.com.sg/retail/news/online-shopping-killing-department-stores-in-singapore

http://ref.data.gov.sg/Agency_Data/NEA/26110600000 [accessed December 2015].

https://www.google.com/maps/d/viewer?mid=zj7DJqcaIyQg.kA-CFlfDoZpk&hl=en_US [accessed December 2015].

http://www.shopping.sg/shopaglore/list-of-shopping-malls-in-singapore/ [accessed January 2016].

32.   The Shopping Trail: Orchard Road

Bartholomew-Clyde (1990). *Singapore City*. 1: 10 000.

Chua Beng Huat (2003), op. cit.

De Koninck (1992), op. cit., p.115.

Singapore Guide and Street Directory (1958).

Singapore Street Atlas 2005-2006. Singapore: Periplus Editions.

Singapore Street Atlas 2015. Singapore: Periplus Editions.

Singapore Street Directory (1988).

Singapore Street Directory (1991).

Singapore Street Directory (2007). Singapore: Mighty Minds.

Singapore Street Directory (2015).

Singapore Tourist Promotion Board (1990). *Singapore Hotels*.

Wildermuth, B. (2015), op. cit.

33.   Hosting Foreign Visitors

Singapore Tourist Promotion Board (1990). *Singapore Hotels*.

Chin K.Y. (2006). "Productivity and efficiency of hotels in Singapore." Thesis, National University of Singapore.

Singapore Guide and Street Directory (1958).

Singapore Street Directory (1993).

Singapore Street Directory (2015).

Singapore Tourist Promotion Board (1990). *Singapore Hotels*.

https://www.stb.gov.sg/ [accessed December 2012].

34.   Taking Care of People's Health

Britnell, Mark (2015). In Search of the Perfect Health System. London: Palgrave.

Most Efficient Health Care, Bloomberg [accessed 3 March 2016].

Singapore Guide and Street Directory (1958).

Singapore Street Directory (1993).

Singapore Street Directory (2015).

Tucci, John (2004). "The Singapore health system: achieving positive outcomes with low expenditure." *Watson Wyatt Healthcare Market Review*, October 2004.

http://www.bloomberg.com/visual-data/best-and-worst//most-efficient-health-care-2014-countries [accessed February 2016].

https://www.moh.gov.sg/index.html [accessed December 2015].

https://ref.data.gov.sg/ [accessed December 2015].

35.   Honouring the Past

Tay Kheng Soon & Akitek Tenggara (1997). *Line, Edge and Shade. The Search for a Design Language in Tropical Asia*. Singapore: Page One Publishing.

https://www.ura.gov.sg/ [accessed December 2015].

http://www.nhb.gov.sg/ [accessed December 2015].

http://www.onemap.sg/index.html [accessed December 2015].

36.   Regulating Outdoor Gastronomy

Photograph by David Berkowitz (Flickr user davidberkowitz) shared on a CC-BY 2.0 license.

Ashar Ghani (2011). *A Recipe for Success: How Singapore Hawker Centres Came to Be*. Institute of Policy Studies.

Kaye, Barrington (1960). *Upper Nanking Street, Singapore*. Singapore: Singapore University Press.

Kong, Lily (2007). *Singapore Hawker Centres. People, Places, Food*. Singapore: National Environment Agency.

Ohtsuka, K., Kinoshita, H., and Marumo, H. (2008). "A Study on the Historical Change of Hawker Centres in Singapore." *J. Archit. Planning*, AIJ, 73, 267: 1029–36.

http://www.nea.gov.sg/ [accessed December 2015].

https://ref.data.gov.sg [accessed December 2015].

http://eresources.nlb.gov.sg/infopedia/articles/SIP_1637_2010-01-31.html [accessed January 2016].

http://www.metropolasia.com/hawker_centres_:_a_truly_singaporean_dining_experience [accessed January 2016].

http://www.nea.gov.sg/public-health/hawker-centres/the-story-of-hawker-centres-upgrading-programme-(hup)# [accessed January 2016].

37. Shifting Electoral boundaries

Turnbull, M.C. (2009), op. cit.

http://www.six-six.com/singapore-elections/

38. Banking on Singapore

Gosh, Abhijit (2015). "Race to be the preferred Asian location." *The Business Times*, 28 January 2015.

MAS, Directory of Banks and Financial Institutions (1975) https://masnetsvc.mas.gov.sg/FID.html [accessed December 2015].

Turnbull, M.C. (2009), op. cit.

http://www.abs.org.sg/aboutus_membership.php [accessed December 2015].

http://www.mas.gov.sg [accessed February 2016].

http://www.singstat.gov.sg/statistics/visualising-data/charts/share-of-gdp-by-industry [accessed April 2016].

39. Doors Wide Open to the World

Chia L.S. (1989). "The Port of Singapore." In Sandhu and Wheatley (eds.), op. cit., pp. 314–36.

De Koninck (1992), op. cit., p. 73.

Google Earth, several dates since 2006.

Maritime and Port Authority. http://www.mpa.gov.sg/

Ministry of Defence, Mapping Unit (1958). Topographic 1: 63,360.

_____ (1987). Topographic 1: 50,000.

Ministry of Defence (2007). *The Air Force*. http://www.mindef.gov.sg/rsaf/

OpenStreetMap (2015).

Port of Singapore Authority. *Annual Reports*.

Port of Singapore Authority. http://www.singaporepsa.com/

Singapore Street Directory (2015).

40. A Harbour City

Chang T.C. (2004). "Tourism in a 'Borderless' World-The Singapore Experience." *AsiaPacific* 73.

Google Earth, 2003 and 2015.

Lee S.W. and Cesar Ducruet. 2009. "Spatial Glocalization in Asia-Pacific Hub Port Cities: A Comparison of Hong Kong and Singapore." *Urban Geography* 30, 2: 162–84.

Singapore Harbour Layout.jpg: http://www.shippingtandy.com/features/singapore/ [accessed February 2016].

https://www.singaporepsa.com/our-business/terminals/future-terminal

http://www.mpa.gov.sg/web/portal/home/port-of-singapore/port-operations

http://www.jp.com.sg/singapore/main-port/port-layout/ [accessed February 2016].

http://ifonlysingaporeans.blogspot.ca/2015/06/pasir-panjang-terminals-35b-expansion.html [accessed February 2016].

http://www.jp.com.sg/about-us/introduction/ [accessed February 2016].

https://www.singaporepsa.com/our-business/terminals [accessed March 2016].

https://en.wikipedia.org/wiki/Jurong_Port [accessed March 2013].

http://databank.worldbank.org/data/home.aspx [accessed March 2016].

https://www.stb.gov.sg/ [accessed March 2016].

41. Changi: An Airport in the Sea

Google Earth, several dates since 2006.

Ministry of Defence, Mapping Unit (1958). Topographic 1: 25,000 (7 sheets).

OpenStreetMap (2015).

Singapore Street Atlas 2005-2006. Singapore: Periplus Editions.

Singapore Street Atlas 2011. Singapore: Periplus Editions.

Singapore Street Directory (2015).

Urban Redevelopment Authority (2003),

http://spring.ura.gov.sg/dcd/eservices/sop/main.cfm?viewmpview

https://www.ura.gov.sg [accessed April 2016].

42. SIA: A World Class Airline and Its Siblings

Map adapted from: De Koninck (2006), op. cit., Map 4, p. 49.

Singapore Airlines Annual Report 2014–2015.

www.singaporeair.com/saa/app/saa [access April 2016].

https://www.singaporeair.com/pdf/Investor-Relations/Annual Report/annualreport1415v1.pdf [accessed April 2016].

http://www.singaporeair.com/en_UK/plan-and-book/where-we-fly/ [accessed March 2016].

http://www.tigerair.com/sg/en/destination_map.php [accessed March 2016].

43. Trading Globally

De Koninck, R. and Comtois, C. (1980). "L'accélération de l'intégration du commerce extérieur de l'ASEAN au marché mondial." Études internationales 11, 1: 43–64.

_____ (1980). "ASEAN's Growing Integration Within World Trade." *Asean Business Quarterly* 4, 3: 27–34.

http://data.un.org/ [accessed February 2016].

44. Foreign Lands for Expansion: The Riau Islands

Bintan. Land Use 1: 180,000. Singapore: Periplus Editions.

Bunnell, T., H.B. Muzaini, and J.D. Sidaway (2006). "Global city frontiers: Singapore's hinterland and the contested socio-political geographies of Bintan, Indonesia." *International Journal of Urban and Regional Research* 31, 1: 3–22.

Cleary, Mark and Goh Kim Chuan (2005). *Environment and Development in the Straits of Malacca*. London: Routledge.

Google Earth, several dates since 2006.

Grundy-Warr, Carl, K. Peachey, and Martin Perry (1999). "Fragmented integration in the Singapore–Indonesian border zone: Southeast Asia's 'Growth Triangle' against the global economy." *International Journal of Urban and Regional Research* 23, 2: 304–28.

Map adapted from: De Koninck (2006), op. cit., Map 8, p. 73.

Priyandes, Alpano and Majid, M. Rafee (2009). "Impact of Reclamation Activities on the Environment Case Study: Reclamation in Northern Coast of Batam." *Jurnal Alam Bina* 15, 1: 21-34.

Riau Islands. Topographic 1: 500,000.

Riau Islands. Land Use 1: 700,000. Singapore: Periplus Editions.

Sparke, M., J.D. Sidaway, T. Bunnell, and C. Grundy-Warr (2004). "Triangulating the borderless world: geographies of power in the Indonesia–Malaysia–Singapore Growth Triangle." *Transactions of the Institute of British Geographers* 29, 4: 485–98.

http://en.tempo.co/read/news/2015/10/07/206707300/Batam-Mangrove-Forest-Shrinking-in-Size [accessed March 2016].

http://m.thejakartapost.com/news/2014/03/06/labor-issues-blamed-investment-slump-batam.html [accessed March 2016].

http://goodelectronics.org/news-en/indonesian-workers-strike-against-philips-union-busting [accessed March 2016].

45. Foreign Lands for Expansion: Johor

Comaroff, Joshua (2015). "Built on Sand; Singapore and the New State of Risk." *Harvard Design Magazine*, 39.

Comprehensive Development Plan for South Johor Economic Region 2006-2025. Kuala Lumpur: Khazanah Nasional, 2006.

Map adapted from: De Koninck, R. (2007), *Malaysia. La dualité territoriale*. Paris: Belin, Map 13, p. 109.

http://www.harvarddesignmagazine.org/issues/39/built-on-sand-singapore-and-the-new-state-of-risk [accessed April 2016].

http://www.thestar.com.my/business/business-news/2013/02/19/land-and-development-on-manmade-danga-bay-island-to-cost-rm8bil/ [accessed April 2016].

http://www.thestar.com.my/business/business-news/2013/02/25/iskandar-waterfront-looks-to-raise-up-to-rm3b-considers-listing-on-msia-and-spore-exchanges/?style=biz [accessed April 2016].

https://en.wikipedia.org/wiki/Iskandar_Malaysia [accessed April 2016].

46. Foreign Lands for Expansion: The World

Map adapted from De Koninck (2006), op. cit., Map 5, pp. 50–1.

http://www.iesingapore.gov.sg/wps/portal

http://www.mpa.gov.sg/

http://welcome.singtel.com/default.asp

http://www.singstat.gov.sg/stats/

http://info.singtel.com/business/singtel-global-offices [accessed April 2016].

https://en.wikipedia.org/wiki/Singtel [accessed April 2016].

http://www.iesingapore.gov.sg/IE%20Corporate/Contact%20Info/Global%20Networks [accessed April 2016].

http://www.embassypages.com/singapore [accessed April 2016].

http://www.singstat.gov.sg/publications/publications-and-papers/investment/singapore's-direct-investment-abroad [accessed April 2016].

http://www.capitalandsingapore.com/ [accessed April 2016].

https://en.wikipedia.org/wiki/Surbana_International_Consultants_Pte_Ltd [accessed April 2016].

http://www.jurong.com/index.php?option=com_content&view=article&id=11&Itemid=29 [accessed April 2016].

47. The World as a Harbour

PSA International annual Report, 2015 https://www.globalpsa.com/ar/ [accessed April 2016].

https://www.globalpsa.com/our-people/ [accessed April 2016].

https://en.wikipedia.org/wiki/PSA_International [accessed April 2016].

http://www.mpa.gov.sg/web/portal/home/port-of-singapore/port-statistics [accessed April 2016].

http://www.mpa.gov.sg/web/portal/home/port-of-singapore [accessed April 2016].

http://www.mpa.gov.sg/web/portal/home/port-of-singapore/port-statistics [accessed April 2016].

https://en.wikipedia.org/wiki/Maritime_and_Port_Authority_of_Singapore [accessed April 2016].

http://www.kepcorp.com/en/content.aspx?sid=80 [accessed April 2016].

48. The World as Military Training Ground

Lee Kuan Yew (2000). *From Third World to First. The Singapore Story: 1965-2000*. New York: HarperCollins.

National Cadet Corps Training Manual, 2016, Singapore.

OpenStreetMap (2015).

Singapore Street Directory (2015).

http://www.mindef.gov.sg/imindef/key_topics/overseas_operations.html [accessed April 2016].

http://www.mindef.gov.sg/imindef/key_topics/overseas_operations/peacesupportops/home.html [accessed April 2016].

http://www.channelnewsasia.com/news/singapore/president-tony-tan-opens/1863612.html [accessed April 2016].

OpenStreetMap (2015).

Singapore Street Directory (2015).

49. From Master Plan to Revised Model

Google Earth, several dates since 2006.

Singapore Street Atlas 2005-2006. Singapore: Periplus Editions.

Singapore Street Atlas 2015. Singapore: Periplus Editions.

Singapore Street Directory (2007). Singapore: Mighty Minds.

Urban Redevelopment Authority (2003),

http://spring.ura.gov.sg/dcd/eservices/sop/main.cfm?viewmpview

50. Following the Plan

Google Earth 2006.

Lee Kuan Yew (2000), op. cit.

Ministry of Trade and Industry Singapore (2003). *New Challenges, Fresh Goals. Towards a Dynamic Global City.*

OpenStreetMap (2015).

Singapore Street Atlas 2005-2006.

Singapore Street Directory (2007). Singapore: Mighty Minds

Singapore Street Directory (2015).

Urban Redevelopment Authority (2003)

http://spring.ura.gov.sg/dcd/eservices/sop/main.cfm?viewmpview

II. Other Useful Sources

(Note: Some of these references have been mentioned in the previous section.)

Architecture of Territory: Hinterland, Singapore, Johor, Riau (2013). ETH Zurich DArch, FCL Singapore.

Architecture of Territory: Sea Region, Singapore, Johor, Riau Archipelago (2014). ETH Zurich DArch, FCL Singapore.

Barnard, Timothy P., ed. (2014). *Nature Contained: Environmental Histories of Singapore.* Singapore: NUS Press.

Beamish, Jane and Jane Ferguson (1985). *A History of Singapore Architecture.* Singapore: Graham Brash.

Brunet, Roger (2001). *Le déchiffrement du monde.* Paris: Belin.

Brunet, Roger and Olivier Dollfus (1990). *Mondes nouveaux,* collection Géographie Universelle, vol. 1. Paris: Hachette / RECLUS.

Buchanan, Iain (1971). *Singapore in Southeast Asia. An Economic and Political Appraisal.* London: Bell and Sons.

Chan Heng Chee (1989). "The Structuring of the Political System." In K.S. Sandhu and Paul Wheatley (eds.), *Management of Success. The Moulding of Modern Singapore.* Singapore: Institute of Southeast Asian Studies, pp. 70–89.

Chia Lin Sien, Asafur Rahman, and Dorothy Tay B.H., eds. (1991). *The Biophysical Environment of Singapore.* Singapore: Singapore University Press.

Chua Beng Huat (1997). *Communitarian Ideology and Democracy in Singapore.* London: Routledge.

_____ (2003). *Life is not Complete without Shopping: Consumption Culture in Singapore.* Singapore: Singapore University Press.

Clammer, John (1985). *Singapore. Ideology, Society, Culture.* Singapore: Chopmen Publishers.

Coless, B.E. (1969). "The Ancient History of Singapore." *Journal of Southeast Asian History* 10, 1: 1–11.

De Koninck, Rodolphe (1975). *Farmers of a City-State: The Chinese Smallholders of Singapore.* Montreal: Canadian Sociology and Anthropology Association.

_____ (1990). "Singapore or the Revolution of Territory. Part One: the Hypothesis." *Cahiers de Géographie du Québec* 92: 209–16.

_____ (1992). *Singapour. Un atlas de la révolution du territoire / Singapore. An Atlas of the Revolution of Territory.* Montpellier: RECLUS

_____ (2006). *Singapour, la Cité-État ambitieuse.* Paris: Belin.

_____ (2007). *Malaysia, la dualité territorial.* Paris: Belin.

_____ (2012). *L'Asie du Sud-Est,* 3rd ed. Paris: Armand Colin.

De Koninck, Rodolphe, Drolet, Julie, and Girard, Marc (2008). *Singapore. An Atlas of Perpetual Territorial Transformation.* Singapore: NUS Press.

Dobby, E.H.G. (1940). "Singapore: Town and Country." *Geographical Review* 30: 84–109.

Edwards, Norman and Peter Keys (1988). *Singapore. A Guide to Buildings, Streets, Places.* Singapore: Times Books International.

Enright, D.J. (1969). *Memoirs of a Mendicant Professor.* London: Chatto and Windus.

Friess, Dan and Oliver, Grahame J.H. (2015). *Dynamic Environments of Singapore.* Singapore: McGraw-Hill.

Gamer Robert E. (1972). *The Politics of Urban Development in Singapore.* Ithaca: Cornell University Press.

George, Cherian (2000). *Singapore. The Air-Conditioned Nation. Essays on the politics of comfort and control.* Singapore: Landmark Books.

Gomez, James (2002). *Internet Politics: Surveillance and Intimidation in Singapore.* Singapore: Think Centre.

Government of Singapore (1991). *Singapore: The Next Lap.* Singapore: Times Editions.

Gupta, Avijit and John Pitts (1992).*The Singapore Story. Physical Adjustments in a Changing Landscape.* Singapore: Singapore University Press.

Haas, Michael, ed. (1999). *The Singapore Puzzle.* London: Praeger.

Hancok, T.H.H. (1986). *Coleman's Singapore.* Kuala Lumpur: Malaysian Branch of the Royal Asiatic Society.

Hodder, Brian W. (1953). "Racial Groupings in Singapore." *Malayan Journal of Tropical Geography* 1: 25–36.

Humphrey, John W. (1985). *Geographic Analysis of Singapore's Population.* Singapore: Department of Statistics.

Koolhas (1995). "Singapore Songlines: Portrait of a Potemkin Metropolis. Or Thirty Years of Tabula Rasa." In Kem Koolhas and Bruce Mau (dir.) *S, M, L, XL.* Monacelli Press.

Koh, Tommy and Koh, Richard W.J. (2015). *Over Singapore.* Singapore: Editions Didier Millet.

Kwa Chong Guan (2004). "From Temasek to Singapore: Locating a Global City-State in the Cycle of Melaka Straits History." In John N. Miksic and Cheryl-Ann Low Mei Gek (eds.), *Early Singapore. 1300–1819.* Singapore History Museum, pp. 124–46.

Lai Chee Kien (2015). *Through the Lens of Lee Kip Lin: Photographs of Singapore 1965-1995.* Singapore: Editions Didier Millet and National Library Board.

Lee Kip Lin (1988). *The Singapore House 1819–1942.* Singapore: Times Edition.

Lefebvre, Henri (1974). *La production de l'espace.* Paris: Anthropos.

_____ (1991). *The Production of Space.* Oxford: Blackwell.

Li, Tania (1989). *Malays in Singapore. Culture, Economy and Ideology.* Singapore: Oxford University Press.

Lim, William S.W., ed. (2002). *Postmodern Singapore.* Singapore, Select Publishing.

_____ (2004). *Architecture, Art, Identity in Singapore. Is there Life after Tabula Rasa?* Singapore: Asian Urban Lab.

Low, Linda, ed. (1999). *Singapore. Towards a Developed Status.* Singapore: Oxford University Press.

MacLeish, Kenneth and Winfield Parks (1966). "Singapore. Reluctant Nation." *National Geographic* 130, 2: 269–300.

Miksic, John N. (2000). "Heterogenetic Cities in Premodern Southeast Asia." *World Archaeology* 32, 1: 106–20.

Miksic, John and Low Cheryl-Ann Mei Gek, eds. (2004). *Early Singapore 1300s–1819.* Singapore: Singapore History Museum.

Miksic, John (2013). *Singapore and the Silk Road of the Sea, 1300–1800.* Singapore: NUS Press.

Ministry of Trade and Industry (1986). *The Singapore Economy. New Directions.* Singapore.

_____ (1991). *The Strategic Economic Plan. Towards a Developed Nation.* Singapore.

_____ (2003). *New Challenges, Fresh Goals. Towards a Dynamic Global City. Report of the Economic Review Committee.* Singapore.

Neville, Warwick (1965). "The Areal Distribution of Population in Singapore." *Journal of Tropical Geography* 20: 16–25.

_____ (1966). "Singapore: Ethnic Diversity and its Implications." *Annals of the Association of American Geographers* 61, 1: 236–53.

_____ (1969). "The Distribution of Population in the Post-War Period." In Ooi Jin Bee and Chiang Hai Ding (eds.), *Modern Singapore.* Singapore: University of Singapore Press, pp. 52–68.

Ooi Giok Ling (2004). *Future of Space. Planning Space and the City.* Singapore: Eastern University Press.

Ooi Giok Ling and Kenson Kwok, eds. (1997). *City and the State. Singapore Built Environment Revisited.* Singapore: Oxford University Press.

Perry, Martin, Kong, Lily, and Brenda S. A. (1997). *Singapore. A Developmental City State.* New York: John Wiley and Sons.

Quah, John S.T., ed. (1990). *In Search of Singapore's National Values.* Singapore: Times Academic Press.

Raban, Jonathan (1974). *Soft City. A Documentary Exploration of Metropolitan Life.* Harvill Press.

Reith, G.M. (1985). *Handbook to Singapore, 1892 (Revised by Walter Makepeace in 1907).* Singapore: Oxford University Press.

Rodan, Garry (1989). *The Political Economy of Singapore's Industrialization. National State and International Capital.* Kuala Lumpur: Forum

_____ (2004). *Transparency and Authoritarian Rule in Southeast Asia. Singapore and Malaysia.* Londres: RoutledgeCurzon.

Raffestin, Claude (1980). *Pour une géographie du pouvoir.* Paris: Librairies techniques.

Raffestin, Claude and Bresso, Mercedes (1979). *Travail, espace, pouvoir.* Geneva: l'Âge d'Homme.

Scott, James (1998). *Seeing Like a State. How Certain Schemes to Improve the Human Condition Have Failed.* Yale University Press.

Siddique, Sharon and Nirmala Pura Shotam (1982). *Singapore's Little India.* Singapore: Institute of Southeast Asian Studies.

Singapore. The Encyclopedia (2006). Singapore: Editions Didier Millet and National Heritage Board.

Sandhu, K.S. and Wheatley, Paul, eds. (1989). *Management of Success. The Moulding of Modern Singapore.* Singapore: Institute of Southeast Asian Studies.

Tamney, Joseph B. (1996). *The Struggle over Singapore's Soul.* Berlin and New York: Walter de Gruyter.

Tan Lee Wah (1975). "Changes in the Distribution of Population of Singapore. 1957–1970." *Journal of Tropical Geography* 40: 53–62.

Tay Kheng Soon (1989a). "The Architecture of Rapid Transformation." In K.S. Sandhu and Paul Wheatley (eds.), *Management of Success. The Moulding of Modern Singapore*. Singapore: Institute of Southeast Asian Studies, pp. 860–78.

_____ (1989b). *Mega-Cities in the Tropics. Towards an Architectural Agenda for the Future*. Singapore: Institute of Southeast Asian Studies.

Tay Kheng Soon and Akitek Tengarra (1997). *Line, Edge and Shade. The Search for a Design Language in Tropical Asia*. Singapore: Page One Publishing.

Teo, Peggy, Brenda S.A. Yeoh, Ooi Giok Ling, and Karen, P.Y. Lai (2004). *Changing Landscapes of Singapore*. Singapore: McGraw Hill.

Teo Siew Eng (1978). "Spatial Patterns of Residential Moves in an Asian City: the Singapore Experience." *Journal of Tropical Geography* 46: 86–94.

Trocki, Carl A. (2006). *Singapore. Wealth, Power and the Culture of Control*. London: Routledge.

Tuan Yi-Fu (1974). *Topophilia. A Study of Environmental Perception, Attitudes and Values*. Englewood Cliffs: Prentice Hall.

Turnbull, C.M. (2009). *A History of Singapore. 1819–2005*. Singapore: NUS Press.

Wee Yeow Chin and Richard Corlett (1986). *The City and the Forest. Plant Life in Urban Singapore*. Singapore: Singapore University Press.

Wheatley, Paul (1954). "Land Use in the Vicinity of Singapore in the eighteen-thirties." *Journal of Tropical Geography* 2: 63–6.

Wong, Aline K. and Ooi Giok Ling (1989). "Spatial Reorganization." In K.S. Sandhu and Paul Wheatley (eds.), *Management of Success. The Moulding of Modern Singapore*. Singapore: Institute of Southeast Asian Studies, pp. 788–812.

Wong Poh Poh (1969a). "The Surface Configuration of Singapore Island: a Quantitative Description." *Journal of Tropical Geography* 29: 64–74.

_____ (1969b). "The Changing Landscapes of Singapore Island." In Ooi Jin Bee and Chiang Hai Ding (eds.), *Modern Singapore*. Singapore: Singapore University Press, pp. 20–51.

_____ (1989) "The Transformation of the Physical Environment." In K.S. Sandhu and Paul Wheatley (eds.), *Management of Success. The Moulding of Modern Singapore*. Singapore: Institute of Southeast Asian Studies, pp. 771–87.

Wong Tai-Chee and Yap Lian-Ho, Adriel (2004). *Four Decades of Transformation. Land Use in Singapore, 1960-2000*. Singapore: Eastern Universities Press.

Yeoh, Brenda S.A. (1996). *Contesting Space, Power Relations and the Urban Built Environment in Colonial Singapore*. Singapore: Oxford University Press.

Yeoh, Brenda S.A. and Shirlena Huang (2004). "Foreign Talent in our Midst: New Challenges to Sense of Community and Ethnic Relations in Singapore." In Lai Ah Eng (ed.), *Beyond Rituals and Riots. Ethnic Pluralism and Social Cohesion in Singapore*. Singapore: Eastern Universities Press, pp. 316–38.

Yeoh, Brenda S.A. and Theresa Wong (2015). *Over Singapore 50 Years Ago. An Aerial View in the 1950s*. Singapore: Editions Didier Millet and National Archives of Singapore

Yeung Yue-Man (1973). *National Development Policy and Urban Transformation in Singapore*. Chicago: University of Chicago.

Useful Websites

www.gov.sg

www.mas.gov.sg/masmcm/bin/pt1Home.htm

www.mpa.gov.sg

www.singstat.gov.sg

www.straitstimes.asia1.com.sg

www.economist.com/countries/Singapore

Index